Thomas W. Wieting

Reed College, 1967

NEUROPHYSIOLOGY:
A Primer

NEUROPHYSIOLOGY:
A Primer

CHARLES F. STEVENS
Department of Physiology and Biophysics
University of Washington School of Medicine

John Wiley & Sons, Inc.,
New York London Sydney

SECOND PRINTING, JUNE, 1967

Library of Congress Catalog Card Number: 66-15872
Printed in the United States of America

PREFACE

This book is intended as a brief, up-to-date introduction to the properties of nerve cells and the interactions between them which are believed to underly the functioning of the nervous system. I have taken special care to organize the presentation in a way which I hope will make most of the material easily accessible to readers without training in biology, physics, or mathematics. My aim has been to give an overview of modern neurophysiology which will indicate to the reader what neurophysiologists are doing and provide him with a framework of basic terminology and empirical observations useful for approaching more specialized and detailed works.

In organizing and selecting the material I had in mind several groups of potential readers. First, of course, are the medical and graduate students who would like to find out what neurophysiology is about before plunging into the lengthier and more comprehensive treatments. Next, I was thinking of psychologists and biochemists who require a brief introduction to modern views of nervous system function. I hope these readers will find that this book makes communication with neurophysiologists somewhat easier. Finally, I intended the book for readers whose training is in the physical sciences—physics, mathematics, or engineering—and whose interests are turning toward the nervous system.

In my experience, there are three main difficulties in learning neurophysiology. First, neurophysiology has so many technical terms (often arising historically) that simply learning the language can be quite difficult. Second, because of the nature of the subject, a fair amount of knowledge from mathematics, biology, physics, and physical chemistry is often required. Finally, the reader is often unable to see the point of what, at times, seems like a series of disconnected observations about nerve cells; he cannot construct a coherent picture from the available facts. Unfortunately, these difficulties are inherent in the subject matter, and it seems to me impossible to eliminate them entirely. However, I have been aware of them, and have tried to present the material in a way which minimizes confusion arising from these sources.

I have used as few technical terms as possible, and have defined them carefully, usually illustrating the term when first introduced. In treating the required anatomy I have presented an accurate, simplified picture of neuron and brain structure without going into more detail than absolutely necessary. To minimize the number of concepts required from mathematics, physics, and physical chemistry, the properties of neurons have been developed from a qualitative and descriptive point of view in early chapters, a discussion of the ionic basis for the action potential being postponed until the end. Among the advantages of this order of presentation is the fact that neuron properties can be introduced entirely in terms of membrane potential without initial concern with the more difficult notions involving current flows.

Systematizing the observations of neurophysiology presents the most difficult problem. I have attempted to give a coherent picture by organizing the presentation around what I have termed the "slow potential" theory, that is, the notion that computations are performed in the nervous system by summing slow potentials constructed from temporal and spatial summations of postsynaptic potentials. A certain amount of unity is thus achieved, but at the risk of providing, by misleading emphasis, a distorted picture of nervous system function. To guard against this, I have indicated to the reader the occasions when this risk occurs.

Chapter 1 presents the structural features of the nervous system which are required for the discussion of physiology which follows. The second chapter introduces the action potential and deals descriptively with the properties of axons, and Chapter 3 describes postsynaptic potentials and dendritic properties. How the properties described in the preceding two chapters can be used by the nervous system to perform calculations is the subject of the fourth chapter. Chapter 5 gives a schematization of receptor physiology and the properties of muscle, and the sixth chapter is devoted to a specific example (*Limulus* eye physiology) which illustrates many of the points made in the preceding five chapters.

The first six chapters, then, describe the properties of neurons and present a scheme of nervous system function which shows how these properties can be used to perform certain calculations. In Chapter 7, I consider the techniques which are used to study ensembles of neurons (ablation, gross recording, and stimulation) and illustrate the kind of results one obtains from using these techniques by reference to data on the hypothalamic control of food intake.

Chapter 8 is devoted to the problem of memory. In this chapter, I discuss the difficulties which arise when one attempts to infer neural properties underlying memory from behavior experiments, and I outline some of the requirements which a proposed mechanism for memory must fulfill.

Chapters 9 and 10 are nearly parallel treatments of the ionic mechanism for the action potential and postsynaptic potentials, Chapter 9 being qualitative and Chapter 10 quantitative. The final chapter is intended primarily for those readers from the physical sciences and may be omitted by others.

I have included an optional chapter which presupposes more than an elementary knowledge of physics, physical chemistry, and mathematics for the following reasons. First, readers from the physical sciences should find such a treatment in some respects clearer and more satisfying than the strictly qualitative arguments made earlier. Second, those interested in neurophysiology who wish to learn more physics and mathematics will find this chapter helpful in indicating some of the areas used in modern neurophysiology, and in illustrating how the various techniques are typically applied. Finally, since the same material is not available in another single source, the last chapter will perhaps continue to be a useful reference section for readers who go further in neurophysiology.

Although selection of material and the emphasis I give to it (along with any errors) are, of course, my own, almost all I present is part of the written or verbal tradition of neurophysiology. Parts of the last chapter are not quite standard, and I can lay claim to a few ideas scattered through the text, but for the bulk of the material I am indebted to my teachers, colleagues, and students.

The only figures reproduced from a published source are from Professor Krieg's reconstruction of the rat hypothalamus (Figures 1-5, 1-6, and 7-1); they are included here with his kind permission. Mr. Walter Eva and the editorial staff at John Wiley and Sons have performed much skillful editing of the manuscript, and I gratefully acknowledge their assistance.

Finally, if this book is a readable one, it is primarily because of the efforts of my wife, Jane. Aside from her more direct help in reading the manuscript, her patience throughout the several years has been inexhaustible; without both, the manuscript would surely never have been completed.

CHARLES F. STEVENS

Seattle, Washington

November 1965

CONTENTS

Chapter 1

THE STRUCTURAL BASIS FOR NERVOUS SYSTEM FUNCTION

In the nervous system structure and function are so closely related that it is difficult to understand one without knowing something about the other. Before beginning a discussion of function, therefore, it will be necessary to develop some basic neuroanatomical concepts and terminology. We shall describe the basic element of the nervous system, the neuron, consider the anatomical relationships between neurons, and discuss the overall scheme of organization to which groups of neurons within a nervous system conform.

The human brain contains a complicated network of perhaps ten billion highly specialized cells called **neurons,** or **nerve cells.** Although no two neurons have exactly the same structure, fortunately they do share certain features that are significant for the neurophysiologist. Like other cells, the neuron is bounded by a thin **membrane** (the plasma membrane) and has a **cell body** (or equivalently **soma**). Projecting from the soma are a number of extensions of the cell known as the **dendrites** and the **axon.** Figure 1-1 illustrates three neighboring cells with important parts labeled. Generally, only a single axon arises from the soma, but, as shown in the figure, this axon may give off side branches and characteristically divides up into a number of smaller branches

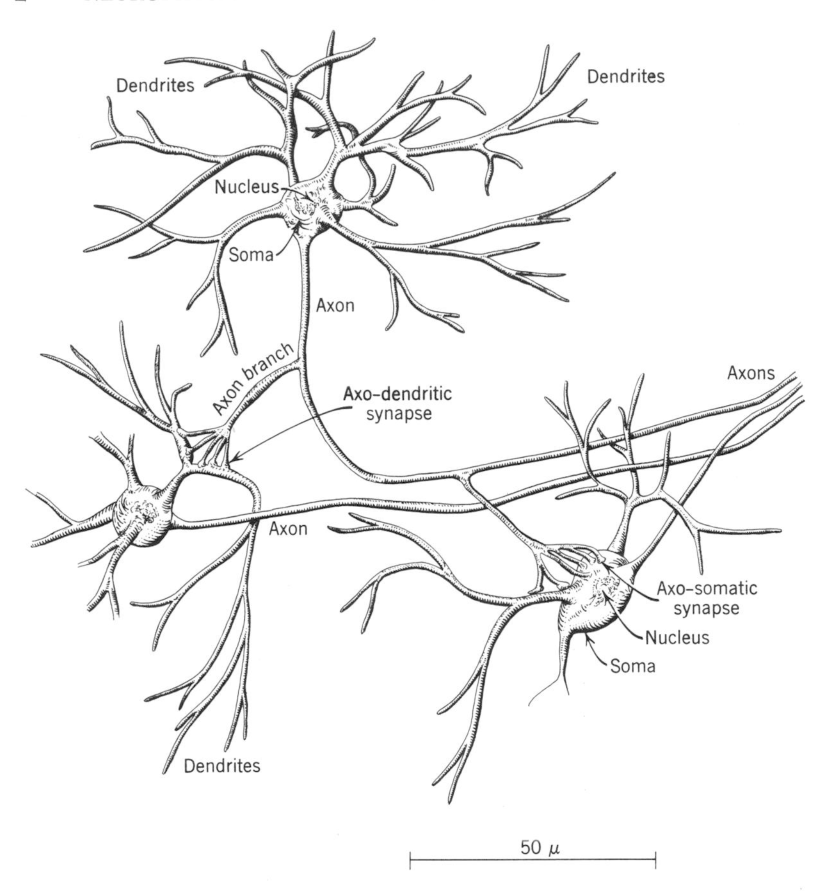

Figure 1-1. Three neurons (semischematic).

just before terminating. As a rule, several or perhaps many dendrites arise from the cell body and then branch again and again to form a complicated treelike structure known, in fact, as the **dendritic tree.** The dimensions of a neuron vary greatly from one to the next, but, to give a general idea of size, the cell body is roughly 30 micra across (1 micron $= 10^{-6}$ meter), and dendrites are perhaps 200 or 300 micra long. It is the length of the axon which is subject to the greatest variation from one nerve cell to the next, for axons may be from perhaps 50 micra to several meters long.

Nerve cells are not isolated but rather are interconnected in a very characteristic way. Special points of contact between neurons, called **synapses,** are of particular importance, for it is at the synapse that information flows from one nerve cell to the next. Also shown in Figure 1-1 is the way in which nerve cells come into contact with each other. Typically, the axon of one nerve cell forms a synapse with (or *synapses with,* using *synapse* as a verb) either the cell body or dendrites of other neurons. Thus there are two principal types of synapses, axo-dendritic (axon with dendrite) and axo-somatic (axon with cell body). In fact, it has been discovered in recent years that virtually all of the surface of a neuron's soma and dendrites is provided with axon terminals, so much so that the somatic and dendritic membranes are literally encrusted with synaptic endings. On the other hand, a synapse between two axons is less frequently observed, although such axo-axonal synapses do occur. Thus the anatomical basis for the main channels of cummunication between neurons is clear: an axon branch from one nerve cell makes synaptic contact with a dendrite or cell body of another neuron.

The synapse is of such central importance that it will be useful to have a more detailed picture of structural features shared by the more common types of synapse. Figure 1-2 shows a schematic cross section through a segment of dendrite and several of the numerous axon terminals which make contact with the dendrite. A narrow space about 200 to 300 Å wide (1 Å = 10^{-10} meter or 10^{-4} micron) separates the presynaptic membrane (the axon terminal adjacent to the dendrite) from the postsynaptic membrane (dendritic part of the synapse); this space is known as the **synaptic cleft.** On **the presynaptic side** are found a number of mitochondria, indicating that the region is metabolically active. Although mitochondria are found in cells of all types, there is a conspicuous structure which is characteristic of synapses, the **synaptic vesicle.** Synaptic vesicles as seen with the electron microscope are circular in cross section, 300 to 600 Å in diameter, and are nearly always found in numbers adjacent to the presynaptic membrane. These structures are so universally found in axon terminals that their presence is one of the principal ways in which synapses are identified by the electron microscopist.

Because neurons have synaptic connections with one another,

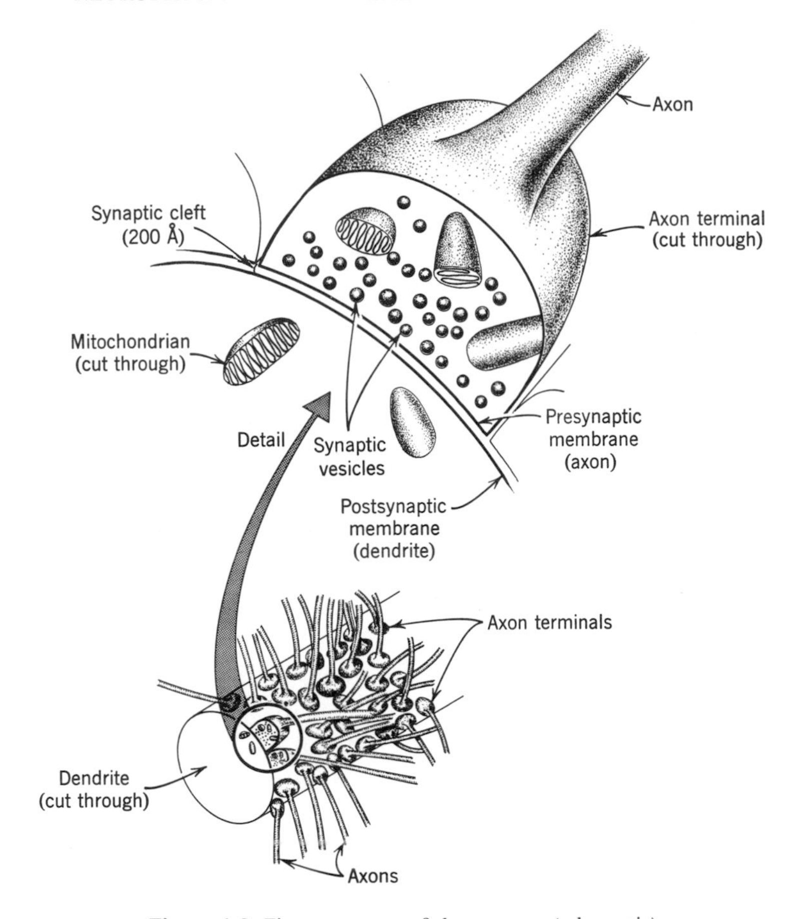

Figure 1-2. Fine structure of the synapse (schematic).

they form neural circuits, and it is the properties of these circuits that determine the behavior of the nervous system and thus of the organism. To give some idea of what the circuits are like, and also to indicate the immense complexity of the brain, an enlarged picture of a thin slice of brain is reproduced in Figure 1-3. A small fraction of the neurons in this brain slice have been stained black. Although direct analysis of such a tangled network is hopeless, a number of special techniques have made it possible to understand such neural circuits to a certain extent.

Although the brain is complicated, fortunately it is not chaotic;

despite individual variations, every brain of a given kind of animal conforms closely to the same scheme of organization. In general, we may distinguish particular brain areas where cell bodies and dendritic trees are concentrated (such areas are frequently called **nuclei**) and other regions which consist primarily of axons running from one group of neurons to another. In these latter regions large numbers of closely packed axons run in parallel to form structures called **tracts,** or alternatively **fiber tracts** (*fiber* here refers to *axon,* for axons are frequently called *fibers* or *nerve fibers*). Similar tracts also run outside of the brain (for example, to muscles), in which case they are rather unfortunately referred to as **nerves.**

Not only are neuron cell bodies grouped to form nuclei, but the nuclei themselves also tend to be clustered together. Thus a single nucleus might contain thousands of neuron cell bodies, and perhaps ten nuclei are clumped together to form a larger structure. Again, the fiber tracts and nuclei are not arranged at random, but according to an orderly scheme characteristic of the type of animal. It is possible to make three-dimensional

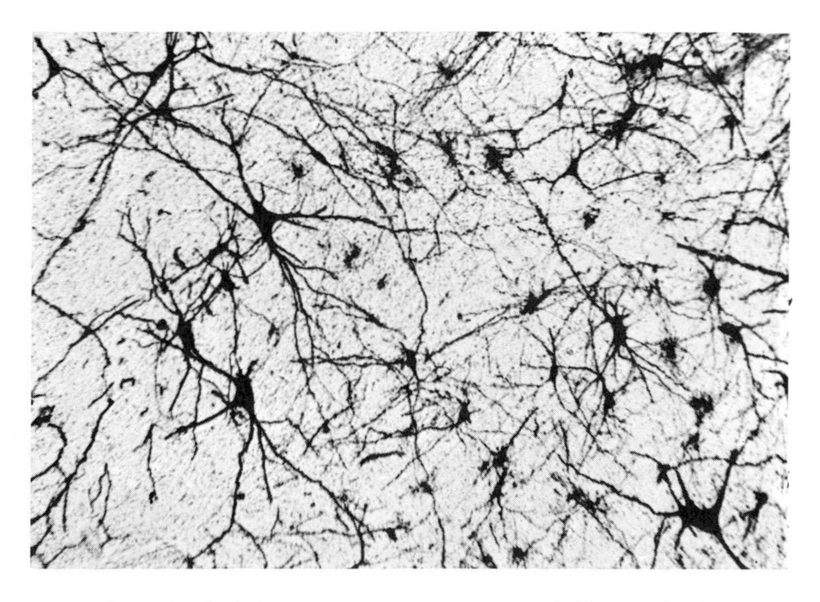

Figure 1-3. Relations between neurons as revealed by Golgi stain.

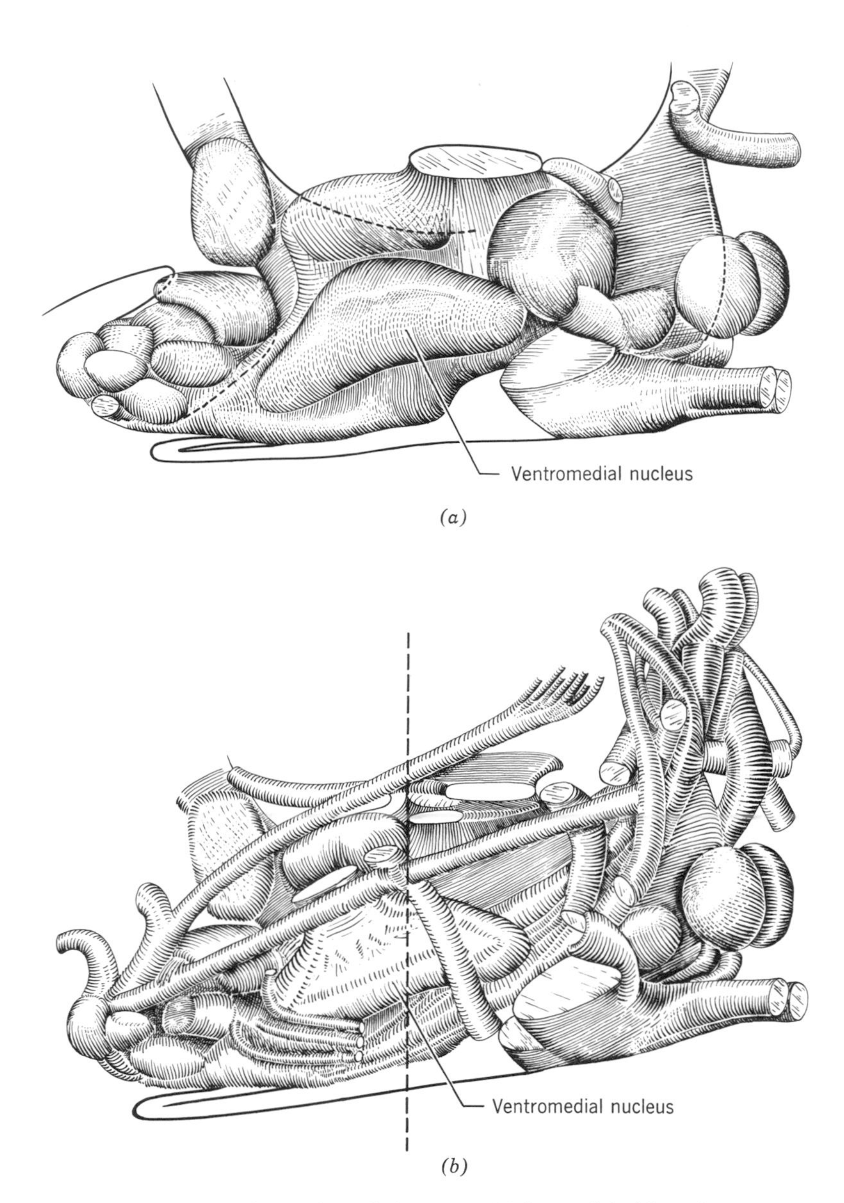

Figure 1-4. Nuclei of the rat hypothalamus. (*a*) Major nuclei without interconnecting fiber tracts. (*b*) Nuclei with some of the major arriving and departing fiber tracts included. (After Krieg, *J. Comp. Neurol.* **55**:19, 1932, Figures 21 and 22.)

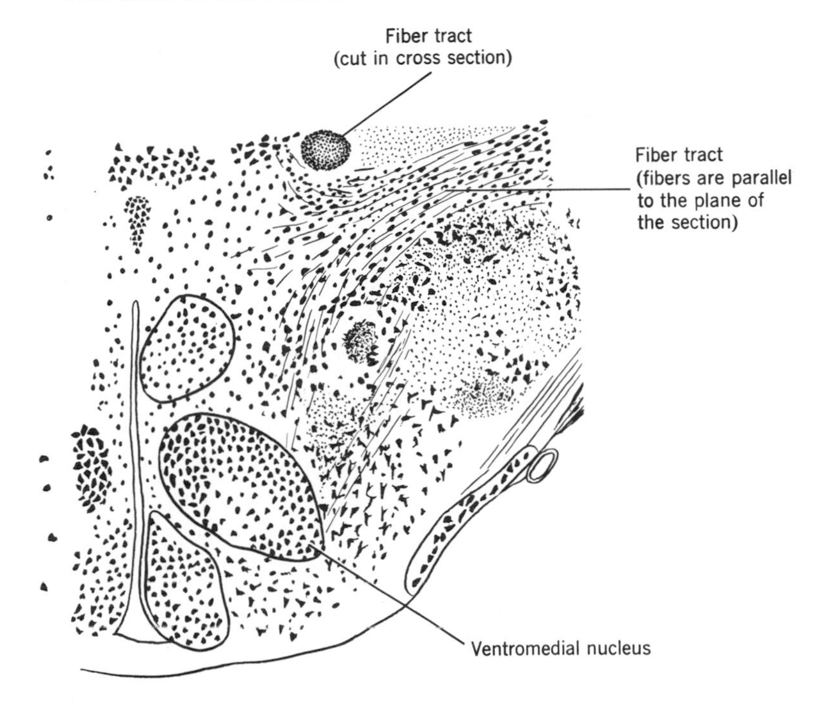

Figure 1-5. Cross section through the rat hypothalamus at level indicated by the dotted line in Figure 1-4*b*. Some of the hypothalamic nuclei are indicated by circles; the black dots within these circles are the cell bodies. (After Krieg, *J. Comp. Neurol.*, **55**:19, 1932, Figure 5.)

maps which give the location and shape of these structures in the brain. In such a map, the different clusters of nuclei are given names like *amygdala* and *thalamus,* and the nuclei within a cluster are designated by terms such as *basolateral* or *ventromedial.* Fiber tracts are also named, either by combining the names of the two parts of the brain connected by the tract (*spinothalamic,* a tract running from the spinal cord to the thalamus) or apparently by whimsey. An example of the sort of map that can be made is shown in Figure 1-4, which illustrates a certain small area of the brain that neuroanatomists call the **hypothalamus.** In Figure 1-5, a cross section through the hypothalamus indicates the location of the nuclei and fiber tracts and shows the size of neuron cell bodies relative to the size of the nuclei.

In addition to neurons, there are, of course, other types of cells in the brain. As elsewhere, we find blood vessels and a certain amount of connective tissue, particularly on the brain surfaces; but the most conspicuous class of cells aside from the neurons is the **neuroglia** or **glia cells.** It is the current view that nearly all the space in the nervous system which is not taken up by the nerve cells themselves is occupied by the glia. There are a number of different types of neuroglia, and their function is still not completely understood; but in general they may be thought of as providing support, structural and metabolic, for the extraordinarily diffuse and delicate neurons. Because the entire functioning of the nervous system depends on countless minute connections between nerve cells, it is vital that the complicated neural circuits have some sort of metabolic protection and mechanical stabilization, a role fulfilled admirably by the glia.

The brain, then, is an immense number of spidery nerve cells, interconnected in a complex net, and embedded in a supporting and protecting meshwork of neuroglia. Dendrites spring from the neuron cell body, branch profusely, and along with the soma receive myriads of axon terminals, making it possible for a single nerve cell to gather information from hundreds of others. Furthermore, cell bodies are collected into groups, the nuclei, and these in turn into clusters. Running back and forth between nuclei or between collections of nuclei are fiber tracts, the main channels of communications between one part of the brain and another. Altogether, these structures are arranged in an orderly way to form the brain of the animal and to provide the anatomical basis for neural function.

Chapter 2

THE AXON: SPECIALIZATION FOR TRANSMISSION OF INFORMATION

The preceding chapter considered the structure of the neuron; this and the succeeding chapters are concerned with neuron function. The neuron is generally considered to be the basic unit of the nervous system—the elementary building block, so to speak—and a detailed understanding of the total nervous system operation thus presupposes a knowledge of how the individual neuron functions. In this and the following two chapters, then, we shall concentrate on certain properties of the individual neuron and on the way these properties can fit into a general scheme of nervous system function.

A neuron receives information through synapses located on its dendrites and soma, integrates this information in its soma, and finally transmits the information over its axon to other nerve cells. Each part of the neuron (axon, dendrite, and soma) has special properties which make it suitable for its particular role in nervous system function. Here we shall consider those properties which specialize the axon for rapidly and reliably transmitting information, and the following two chapters will deal with properties of dendrites and with an overall view of how the neuron integrates information received simultaneously through many different synapses.

First, then, we shall investigate the ways axons are specialized for conducting information from one place to another within the brain. Leaving all details aside for the moment, we may state the final conclusion of these considerations quite briefly: information is transmitted by nerve impulses which travel rapidly along the axon, this information generally being carried as the frequency or number of impulses per second traveling along the axon. Thus the following discussion will center on the properties of axons which give rise to the nerve impulse and to the coding of information.

The nerve impulse is a complicated physico-chemical event. One component of this event is a rapid voltage change which will be described in detail later. Since this voltage change, known as the **action potential,** is actually used in experiments to indicate the presence of a nerve impulse, we shall generally not make a distinction between **nerve impulse** and its component *action potential* in the following discussion.* As a preliminary to discussing the voltage changes which constitute the action potential, we must investigate the electrical properties of the axon. First, however, it is necessary to define four central terms: **membrane potential, resting potential, hyperpolarization,** and **depolarization.**

Because the presence of a nerve impulse is signaled by a rapid fluctuation of the voltage within the axon, much of our discussion will be concerned with such voltage changes. For this reason it is convenient, as well as traditional, to have a special term for the voltage inside the axon, or more precisely, for the voltage difference between the inside and outside of the axon. Since the axon's membrane separates the inside from the outside, this inside-outside voltage difference is referred to as the **membrane potential.** In the new terminology, the presence of a nerve impulse is signaled by a fluctuation in membrane potential.

The first question that might be asked about an axon's mem-

* There is also a somewhat deeper reason for not distinguishing between the nerve impulse and its electrical concomitants: the electrical events associated with the nerve impulse are so intimately involved with the mechanisms through which the impulse occurs that it is difficult to separate the two.

brane potential is: what is the value of the membrane potential when the axon is at rest, that is, when no nerve impulses are occurring? At first thought one might suspect that the resting membrane potential, or simply **resting potential,** as it is usually called, would be zero, for there is no very obvious a priori reason to expect a difference in voltage between the inside and outside of an axon which is at rest. That this is not the case is shown by an experiment such as the following. Suppose that a small electrode is advanced toward a length of axon until it enters the interior by penetrating the axon membrane (Figure 2-1). If the voltage difference between this electrode and the external bathing solution is measured throughout the procedure (Figure 2-2), this voltage will suddenly change from zero to nearly a tenth of a volt negative at the moment the electrode enters the axon. In the resting axon, then, the membrane potential is not zero, but rather a fraction of a volt; the axon is literally a battery with its negative terminal on the inside. If the resting potential were measured over a long period, it would have been found to be constant over time, and if the electrode had penetrated the axon

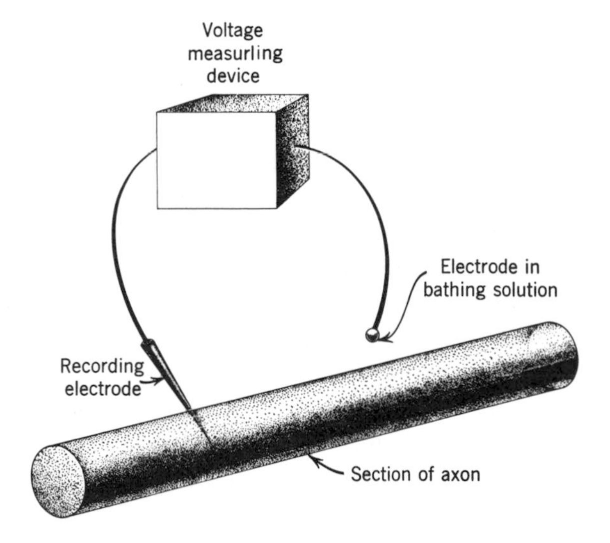

Figure 2-1. Experimental arrangement for recording axon membrane potentials. The axon is surrounded by a salt solution not indicated in the figure.

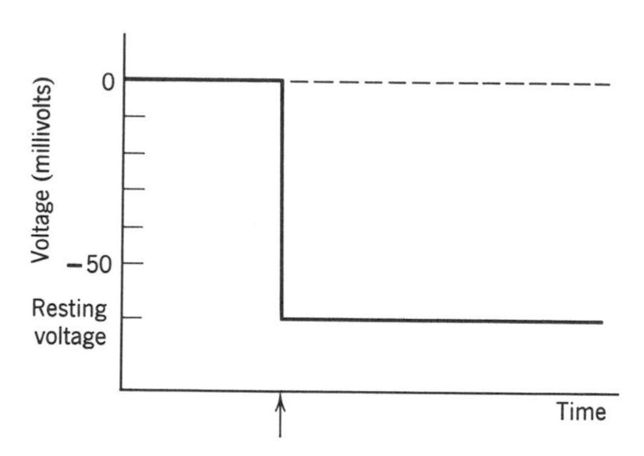

Figure 2-2. Voltage measured by recording electrode as it penetrates an axon. The electrode entered the axon at the time indicated by the arrow.

at a different location, the same value of the resting potential would have been recorded. Furthermore, when the resting potential is measured in different axons from the same animal, or even in axons from different kinds of animals, it characteristically is about the same. Although accurate measurements of resting potential are difficult for various technical reasons, −60 millivolts (1 millivolt = 10^{-3} volt) may be taken as a typical value.

Much of neurophysiology is concerned with variations in the membrane potential, and it will therefore prove convenient to have special terminology with which to describe the direction of these membrane potential fluctuations. Since the axon's membrane potential remains at or near its resting value a large part of the time, the resting potential provides a useful reference point; by convention all fluctuations in membrane potential are compared to the resting potential, which is arbitrarily taken as zero. Whenever the membrane potential is more positive than the resting potential, that is, whenever the inside-outside voltage difference is less than the −60 millivolts seen in the resting state, the axon is said to be **depolarized.** For example, if the membrane potential were −40 millivolts instead of the usual −60 millivolts, the axon would be 20 millivolts depolarized. If, however, the membrane potential is below the resting value, that is, more negative than −60 millivolts, the axon is said to be **hyperpolarized;** an axon with a membrane potential of −70 millivolts would be hyperpolarized by 10 millivolts. Finally, any change which decreases the inside-outside potential difference is known as a depolarization, whereas an increase in the membrane potential is called a hyperpolarization. These terms have their origin in the fact that the axon is normally "polarized," that is, has a resting potential, and alterations in the membrane potential which diminish this resting polarization are *de*polarizations, whereas those acting to increase the polarization are *hyper*polarizations.

Having developed the basic concepts of membrane potential, resting potential, depolarization, and hyperpolarization, we can now present the properties of axons. At the outset, it was noted that the axon is specialized for the transmission of information and that information transfer occurs by nerve impulses sweeping along the axon. The following discussion of properties which specialize the axon for its role as a communication line will be divided into two major sections. The first section will deal with axonal properties which give rise to the nerve impulse, and the second will describe how nerve impulses travel along the axon to transfer information from one place to another.

THE ACTION POTENTIAL AND FREQUENCY CODING

By inserting an electrode into an axon, and by connecting the electrode to an appropriate voltage source (a battery would be the simplest source), the experimenter is able to depolarize the axon to any desired degree. This voltage applied to the electrode is known as the **stimulus.** If a brief stimulus is applied to the axon, a stimulus resulting in a perhaps 15 millivolt depolarization, the response of the axon is an **action potential.** This action potential consists of a brief (approximately ½ millisecond) fluctuation in membrane potential with a characteristic configuration: the membrane potential undergoes a large, rapid depolarization followed by a slightly less rapid return to the resting potential (Figure 2-3). The peak depolarization reached exceeds the magnitude of the resting potential, which means that the axon, usually negative inside, becomes briefly positive on the inside. That is, the depolarization is so large at the peak of the action

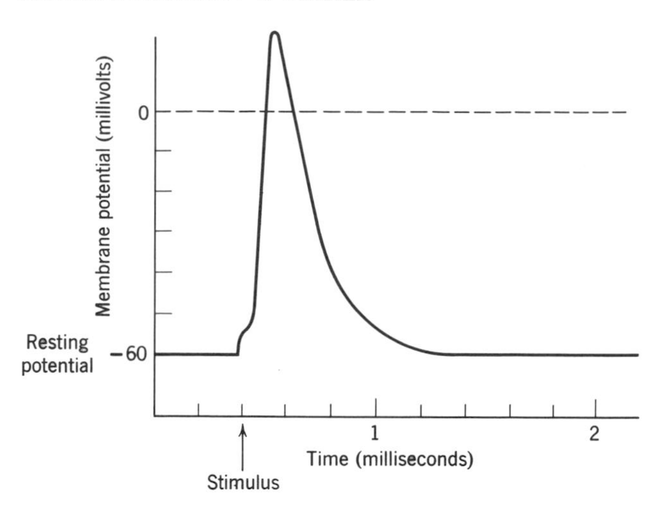

Figure 2-3. An action potential. Stimulus was applied at time indicated by the arrow.

potential that the sign of inside-outside potential difference actually reverses from negative to positive.

The experiment just described suggests a number of questions. What would be the effect of a larger or smaller stimulus? What would happen if the stimulus resulted in a hyperpolarization rather than a depolarization? Is the duration of the stimulus of any importance? What is the effect of two closely spaced stimuli? We shall now investigate these questions and in the process describe properties which specialize the axon for the transmission of information.

In the preceding example a depolarizing stimulus resulted in an action potential; we now consider the response of the axon to a hyperpolarization of the same magnitude (but of opposite polarity, of course). As before, suppose that the electrode has been inserted into the interior of the axon, and suppose further that an appropriate voltage source, such as a battery and a switch, is available for hyperpolarizing the axon to the desired degree. When such a hyperpolarizing stimulus is applied through the electrode, no action potential results, but rather the membrane potential simply follows the stimulus with some slight distor-

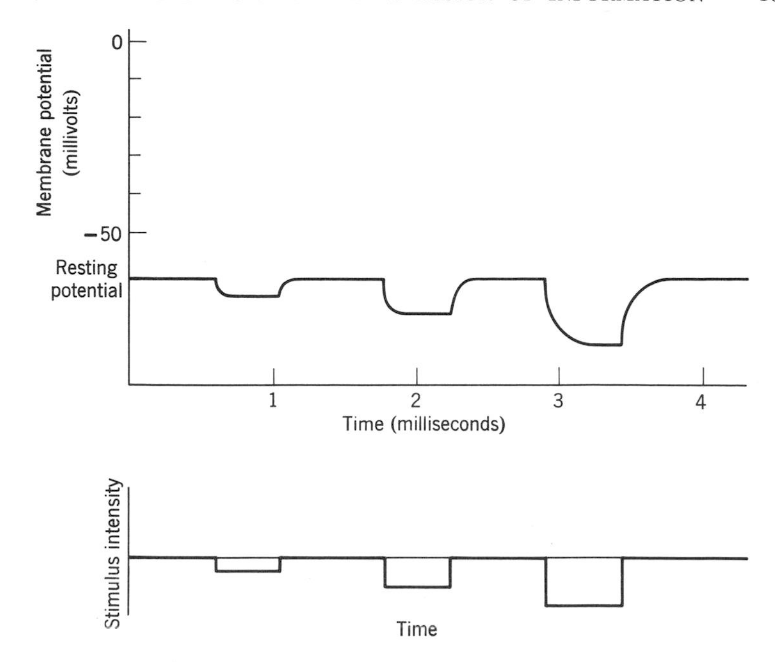

Figure 2-4. Response to hyperpolarizing stimuli of different magnitudes.

tion * (Figure 2-4). If larger or smaller hyperpolarizing stimuli are used, the result is always the same: the membrane potential mirrors, with a slight rounding of sharp corners, the stimulus applied. Because the axon's response to a hyperpolarization is simply a reflection of the stimulus, we say that hyperpolarizing stimuli result in a *passive response* to distinguish between this and the *active response* to depolarization, the action potential. The slight distortion of the hyperpolarizing stimulus results from the *passive* properties of the axon membrane. We note in passing (we shall return to this point in Chapter 10) that the passive properties of the axon membrane are approximately those of a parallel resistance-capacitance circuit and that the preceding has been a description of such a circuit's response to rectangular current pulses.

* This "distortion" is a consequence of the physical properties of the axon membrane, and will be discussed in Chapters 9 and 10.

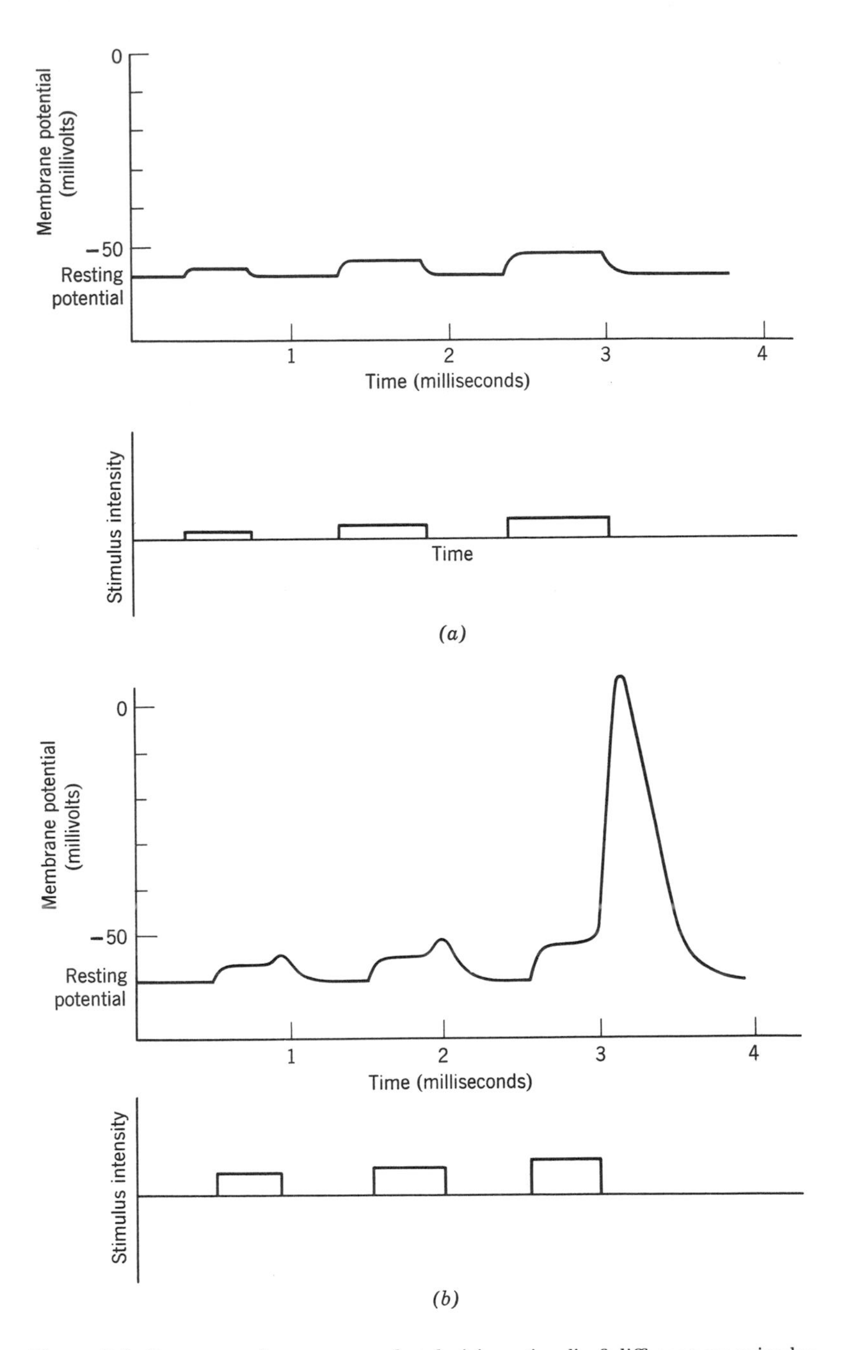

Figure 2-5. Response of an axon to depolarizing stimuli of different magnitudes. (*a*) Small stimuli well below threshold. (*b*) Two stimuli just below threshold for an action potential and one just above threshold. (*c*) Three stimuli well above threshold.

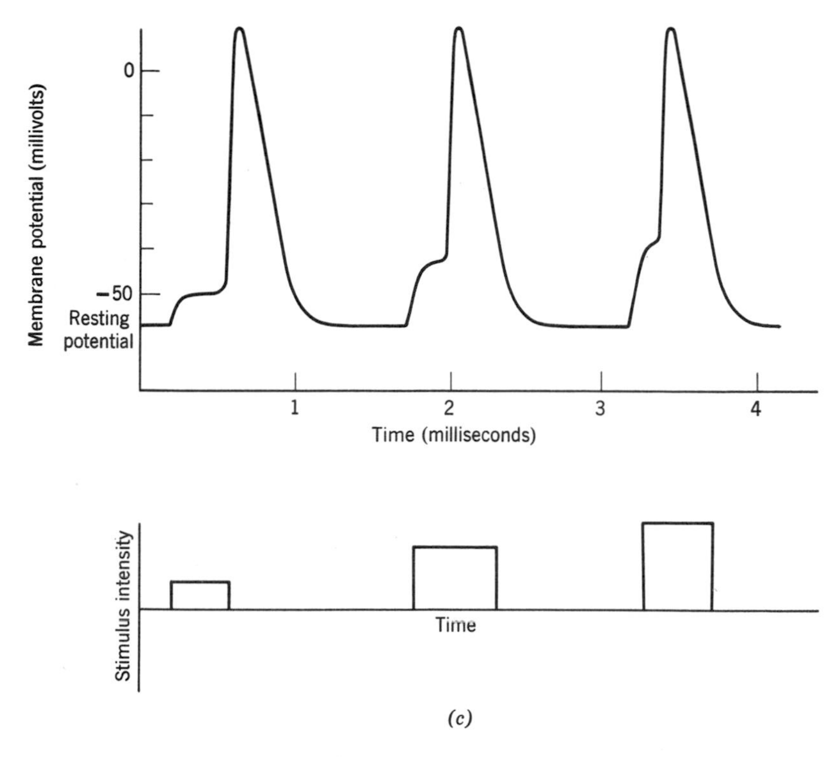

(c)

Figure 2-5 (*continued*).

We turn now to the effect of the magnitude of a depolarizing stimulus on the response of the axon. Suppose that a series of brief depolarizing stimuli of increasing magnitude is presented to an axon through the intracellular electrode. From the preceding description (page 13), we know that at least one magnitude of depolarization results in an action potential; the point of interest is how the axon responds to larger and smaller depolarizations. In such an experiment (Figure 2-5*a*), it is found that, for very small depolarizing stimuli, only a passive response results: the response simply mirrors the stimulus with slight distortion. This passive response is identical to the one seen with hyperpolarizing stimuli, except, of course, that the direction is reversed to mirror the depolarizing stimuli applied. As the intensity of the stimulus is gradually increased, the magnitude of the passive response increases proportionally. With still further increases in stimulus strength, a new active component of the axon's response appears: when the stimulus is removed, a small bump is seen on the falling phase of the passive response (Figure 2-5*b*). As the stimulus magnitude is made still larger, this bump grows in size. Finally, when the stimulus intensity is increased only very slightly more, an action potential of the type illustrated in Figure 2-3 results. Further increases in the stimulus intensity produce no further increase in the size of the axon's response; all action potentials are the same size irrespective of the stimulus magnitude which produced them (Figure 2-5*c*).

Several points are illustrated by the hypothetical experiment just described. First, it is important to note that for small depolarizing stimuli, the axon shows only a passive response. With somewhat stronger stimuli the axon's response deviates slightly from the purely passive response, and a small bump, known for reasons which will become apparent later in the chapter as the local response, appears. As the stimulus intensity is increased on succesive stimulus presentations, a critical stimulus strength is found below which only the local response is seen, and above which an action potential occurs. This critical stimulus strength is known as the **threshold.** Finally, it must be emphasized that any stimulus which results in an action potential produces the same sized action potential irrespective of the magnitude of the stimulus.* This property of the action potential is known as the **all-or-none law** and will be seen a little later in the chapter to have considerable significance.

The all-or-none law states that the size of the action potential is independent of stimulus magnitude; this does not mean, however, that changes in stimulus magnitude are entirely without effect. Suppose that an axon is presented a sequence of stimuli

* Strictly speaking, this statement is misleading in two respects. First, to say that action potentials are always the same size is inaccurate because a variety of conditions are known to alter the action potential amplitude. For example, action potentials occurring during the relative refractory period (see page 20) are typically smaller than those occurring in a rested neuron. Second, in some instances, action potential amplitude does reflect stimulus intensity, even for a nerve impulse traveling along the axon. Nevertheless, neither of these "exceptions" to the all-or-none "law" as given in the text will be important for our purposes.

of increasing intensity and that the time from the onset of the stimulus to the peak of the action potential is measured. Such an experiment reveals (Figure 2-5*c*) that the interval between the onset of the stimulus and the peak of the action potential, this interval is known as the **latency,** decreases systematically as stimulus magnitude increases; the stronger the stimulus, the shorter the latency. A graph of latency as a function of stimulus strength (Figure 2-6) is known as a **strength-latency curve;** it has approximately the same shape for different types of axons although the range of latencies observed may be very different.

Before discussing the significance of preceding observations, it is necessary to present the results of one other type of experiment. Suppose that a stimulus is presented which evokes an action potential and that a second stimulus is presented at various times after the peak of the action potential. How is the threshold affected by the occurrence of an action potential? Such experiments show that immediately after an action potential it is impossible to evoke a second action potential no matter how strong a stimulus is used, and that the threshold returns to normal over a period of perhaps several milliseconds (the time varies greatly with different axons), as compared with the half millisecond duration of the action potential itself (Figure 2-7). The period during which it is impossible to evoke a second action potential is known as the **absolute refractory period,** whereas the longer interval

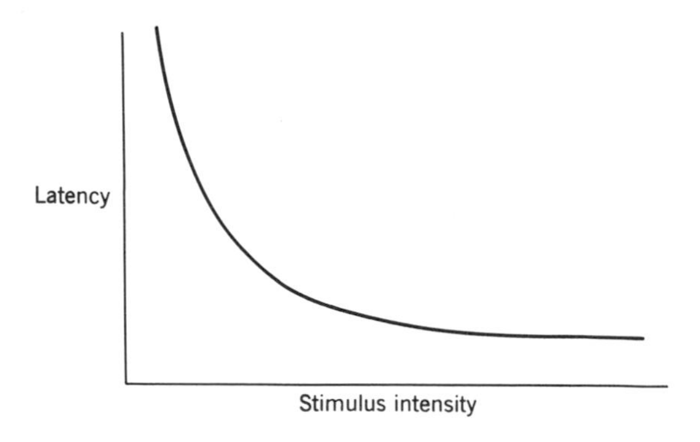

Figure 2-6. Strength-latency curve for an axon.

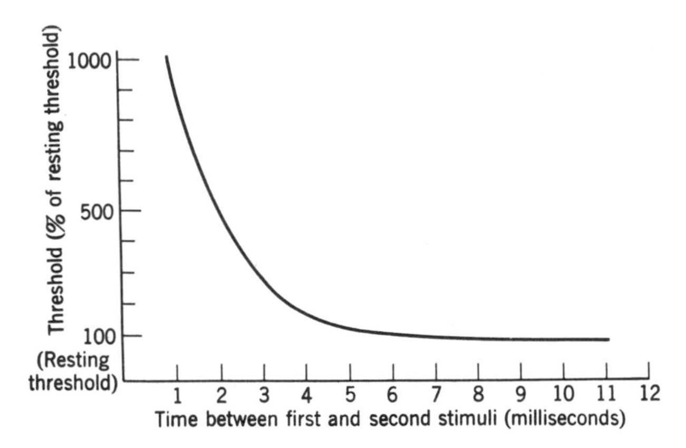

Figure 2-7. Threshold stimulus magnitude as a function of time after an above-threshold stimulus.

during which a stronger than normal stimulus is required is known as the **relative refractory period.**

The important properties of the action potential that have been introduced to this point are the all-or-none law, the threshold, the strength-latency relation, and the refractory period (or more precisely, the return to resting value of the threshold after an action potential). The all-or-none law and the existence of a threshold are not distinct properties, but rather are quite closely related. Specifically, if a sufficiently small stimulus causes no active response (zero stimulus, if necessary), and if all action potentials are the same size irrespective of the magnitude of the stimulus, it must be that there is some critical stimulus intensity below which no action potential occurs, and above which an action potential does occur; this critical stimulus intensity is just the threshold. (The preceding assumes a never decreasing relationship between size of stimulus and response.) The significance of the all-or-none law to the functioning of the nervous system cannot be fully appreciated until the transmission of nerve impulses along the axon is discussed later in the chapter; but at this point it can be seen that since the size of an action potential does not vary, some other property of the action potential must be responsible for carrying information. As previously noted, this property is thought usually to be the rate at which nerve impulses occur.

From the strength-latency curve and the existence of a refractory period, we can see one way in which **frequency coding,** the coding of information in terms of nerve impulse frequency, can arise. Suppose that a long duration stimulus is applied instead of the brief pulselike depolarizations which have been used in the examples discussed. If a long duration *sub-threshold* stimulus is used, it will simply produce depolarization lasting for the period of time that the stimulus was applied (Figure 2-8*a*). However, if a long lasting *above-threshold* stimulus is used, one of several types of behavior can be seen, depending upon the characteristics of the axon. In any axon type an action potential will, of course, occur just after the onset of the stimulus as we have seen previously. Immediately after the first action potential the situation is complicated by the presence of a stimulus during the absolute refractory period when another action potential is not possible. In one type of axon the threshold returns from its high value

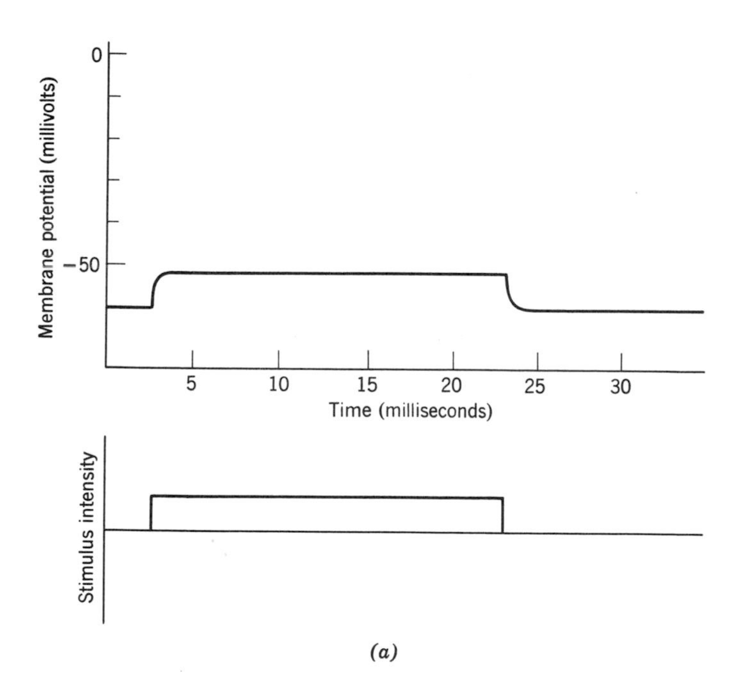

Figure 2-8. Response of an axon to long-lasting depolarizing stimulus. (*a*) Axon response to a below-threshold stimulus. (*b*) Axon response to an above-threshold stimulus. Note that the time scale is different from that in preceding figures.

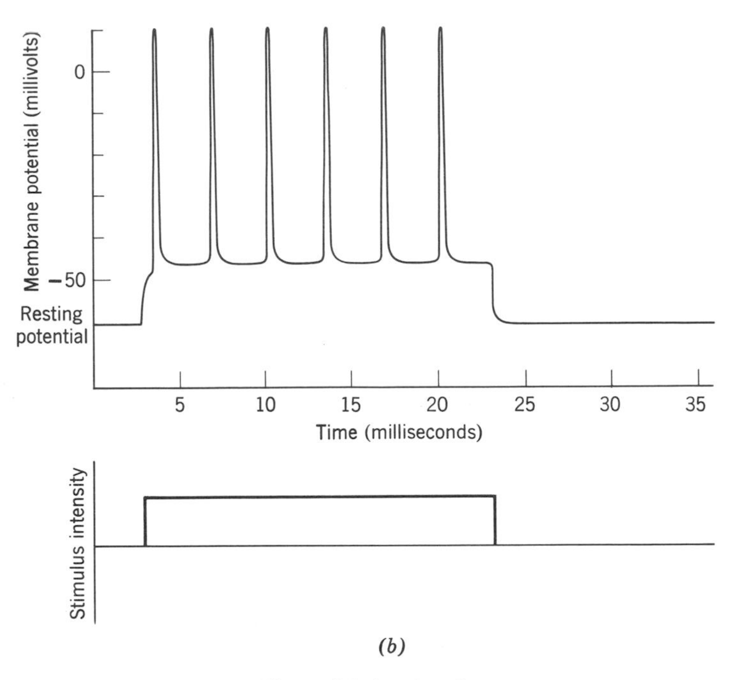

(b)

Figure 2-8 (*continued*).

to the resting level and eventually reaches a point at which the maintained stimulus is effective in causing a second action potential. This same process will reoccur again and again as long as the stimulus is maintained, repeatedly giving rise to action potentials.

By similar reasoning, it is possible to deduce the relationship between nerve impulse frequency and stimulus intensity for the type of neuron we are considering. The repetitive occurrence of nerve impulses in response to a maintained depolarization was predicted in the preceding paragraph on the basis of the refractory period: after one action potential has occurred, a certain amount of time is necessary for the threshold to return to a level where the maintained stimulus is again able to evoke a second action potential. If a stronger stimulus were used, however, there would be a shorter interval between successive impulses since the threshold would not have to return to as low a level for the larger stimulus to be effective. Thus nerve impulse frequency would increase as the magnitude of a maintained depolarization increased.

The strength-latency relation provides a second reason why increasing stimulus strength produces a higher nerve impulse frequency. For a brief stimulus, increasing stimulus magnitude decreases the latency of the action potential (see pp. 18–19, and Figure 2-6). Returning to a maintained stimulus, a stronger stimulus would again be expected to result in a higher frequency of nerve impulses because there would be a shorter time (once a stimulus became effective again after the preceding action potential) before another action potential would occur. The refractory period determines the interval over which a maintained stimulus is ineffective following an action potential, and the strength-latency relation governs the latency of the action potential once the stimulus is again effective. Therefore on the basis of latency and threshold, the larger the stimulus, the higher the frequency at which action potentials would be generated.

In summary, then, the phenomenon of frequency coding is a consequence of several properties of the action potential. The all-or-none law, which states that the size of an action potential is independent of the magnitude of the stimulus which produced it, means that information about stimulus intensity must be carried in some way other than response magnitude. An alternative means of coding information is provided by the properties of refractory period and the strength-latency relation: information about stimulus intensity is contained in the number of nerve impulses occurring per second. The exact relation between stimulus intensity and impulse frequency must of course be determined from more detailed and quantitative investigations; such studies have shown the relation to be linear over an appreciable range of stimulus magnitudes in various types of neurons (Figure 2-9). As might be anticipated from the fact that the strength-latency curve and the return of threshold to its resting value following an impulse occur on quite different time scales in different axons, an axon's frequency for, say, a depolarization 10 millivolts above threshold varies considerably from axon to axon: such a depolarization might cause one axon to produce nerve impulses at a rate of ten per second while the same depolarization would cause a frequency of 500 impulses per second in another axon.

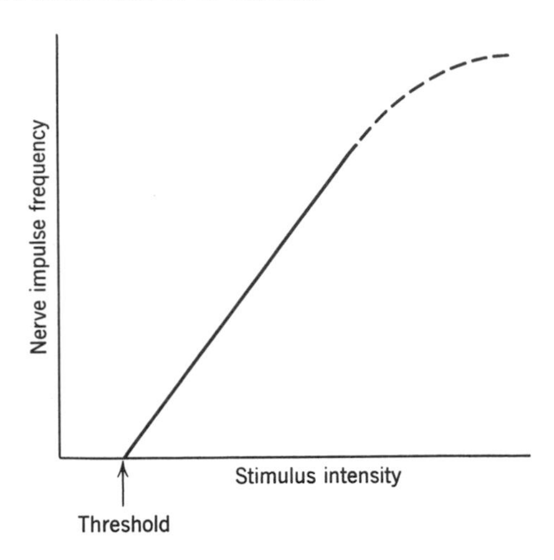

Figure 2-9. Nerve impulse frequency as a function of stimulus intensity for a long-lasting stimulus.

Not all axons respond to a maintained stimulus in the manner described above; that is, a second type of axon does not produce action potentials repetitively when it is subjected to a constant depolarization, but rather produces only one or several action potentials at the onset of the stimulus and then becomes silent in spite of the continuing depolarization.* This phenomenon, usually called **accommodation,** reveals the very approximate nature of the preceding discussion of frequency coding. Further comments on this subject will be made on page 61 of Chapter 4. Although it is reasonably accurate to say that the same properties of the axonal membrane which give rise to frequency coding are also reflected in the strength-latency relation and the refractory periods, the argument of the preceding paragraphs involves an oversimplification in the tacit assumption that the return to threshold following a nerve impulse is the same whether a stimulus

* It should for completeness be pointed out that even axons which produce action potentials repetitively in response to a constant depolarization often exhibit this type of behavior only over a part of their length. Usually only the initial segment near the cell body translates depolarization into nerve impulse frequency. However, since this is the only part of the axon normally subjected to a constant depolarization (why this is true will become clear in later chapters) we have classified axonal properties according to the behavior of this initial segment.

is continuously present or not, and that the strength-latency curve is unaffected by the immediately preceding nerve impulse. In those axons showing marked accommodation, the threshold does not return to normal after an action potential if a continuing depolarization is present, but rather remains so high that the stimulus is insufficient to cause further nerve impulses. In order for another action potential to occur, it is necessary that the depolarization be further increased, or that it be temporarily decreased and then returned to its former level. For this reason, the type of axon showing complete accommodation cannot send information about a constant depolarization, but rather signals *changes* in stimulus strength rather than its absolute magnitude. Both types of axons are present in the central nervous system, although many neurophysiologists feel that the type which can transmit information about the amplitude of a constant stimulus is perhaps more common. Because of their greater simplicity, we shall be most often concerned with axons which exhibit frequency coding of stimulus magnitude.

CONDUCTION OF NERVE IMPULSES ALONG THE AXON

We turn now to the question of how nerve impulses spread down the axon. The problem facing the axon is how to transmit information about stimulus intensity over relatively long distances. Frequency coding together with the propagation of the action potential down the axon will prove to be an admirable solution to this problem. The key to the mechanism of this impulse propagation lies in considering the all-or-none law together with the phenomenon of **passive spread,** a property which we have not previously discussed.

To understand passive spread, we begin by inquiring into the effect of hyperpolarization at one point of the axon on the membrane potential in neighboring regions. Here the experiment described on page 14 is repeated, except that the membrane potential is measured at some distance from the stimulating electrode (that is the electrode connected to a device for increasing

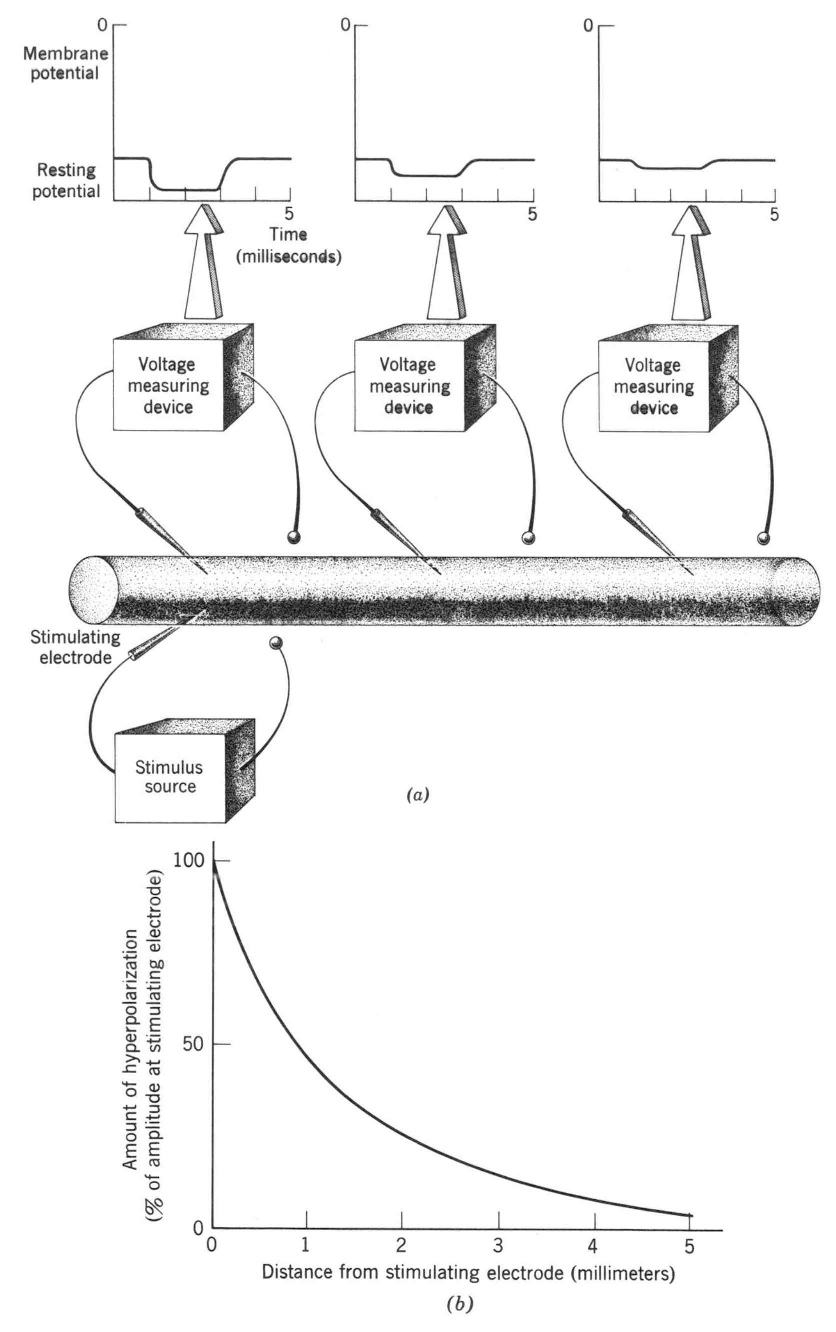

Figure 2-10. Amount of hyperpolarization as a function of distance from a stimulating electrode through which a hyperpolarizing stimulus is applied. (*a*) Experimental arrangement and typical records obtained from three electrodes. (*b*) Typical curve for passive spread in an axon.

membrane potential; see Figure 2-10*a*). In this manner, it has been found that, when one point in the axon is hyperpolarized, neighboring regions are simultaneously hyperpolarized as well, although the effect decreases with distance from the hyperpolarizing electrode (Figure 2-10). Thus if the axon is hyperpolarized by 10 millivolts at one point, half a millimeter away it might be hyperpolarized by 5 millivolts, and 1 millimeter away by 2½ millivolts. If a larger or smaller hyperpolarizing stimulus were used, the results would be analogous: the amount of hyperpolarization would decrease, for example, by half for each half millimeter of distance away from the hyperpolarizing electrode. This phenomenon of a hyperpolarization at one point causing hyperpolarizations in neighboring regions is termed **passive spread** of potential (it is also frequently called **electrotonic spread,** although that term will not be used here). The word *spread* is used because a hyperpolarization at one point may be thought of as spreading to a nearby point, and *passive* is used to distinguish this phenomenon from the active propagation of action potentials which is to be described later in the chapter. *Passive* is chosen rather than some other adjective because all the voltage changes seen in passive spread are only the result of energy supplied by the hyperpolarizing electrode and could just as well be seen in a dead nerve or an inanimate cable with a similar electrical structure.

If the preceding experiment were repeated with depolarizing rather than hyperpolarizing stimuli, the results would be identical as long as very small depolarizations were used. If the complications of local response and action potentials are avoided, it is found that depolarizations spread passively to neighboring regions just as do hyperpolarizations. Clearly once a depolarization is sufficiently large to cause an action potential the situation is not as simple, if only because all stimuli above threshold result in the same sized response. We shall see in the following discussion that the active properties of the axon membrane which lead to the production of action potentials also result in the propagation of the action potential along the axon, so that the action potential in effect spreads down the axon without decreasing in size. Roughly speaking this occurs by the action potential at one location spreading passively to a neighboring region where the depolarization serves as a stimulus to produce a new full-sized action potential. Before investigating this process a little more thoroughly, however, it will be helpful to discuss a variation of the structure of axons, the **myelinated axon.**

A number of axons, the smallest ones in vertebrates, for example, are of the sort already described; they are long thin cylinders. However, many other axons are not simple cylinders, but more complex structures with an overall cylindrical shape interrupted at intervals of about a millimeter by short lengths of axon which are much narrower. The structure of these myelinated axons is basically that of a thin cylinder, the axon proper, encased in a thicker cylindrical sheath, the myelin, which completely surrounds the axon membrane except at the **nodes of Ranvier** where the axon is uncovered for a short segment (Figure 2-11).

Since the axon membrane is not covered at the nodes of Ranvier, it would be expected that here the properties of the axon are as described previously, and this is in fact the case. Nodes show action potentials like those found in unmyelinated axons. However, the myelinated portions of these axons have quite different properties, for they are not thought to give action potentials when depolarized. In the myelinated regions between nodes, the size of both depolarizing and hyperpolarizing responses is proportional to the stimulus intensity and they spread passively along the interior of the axon, decreasing with distance in the characteristic fashion. Because of the properties of the myelin, however, potentials are less attenuated during passive spread than in unmyelinated axons.

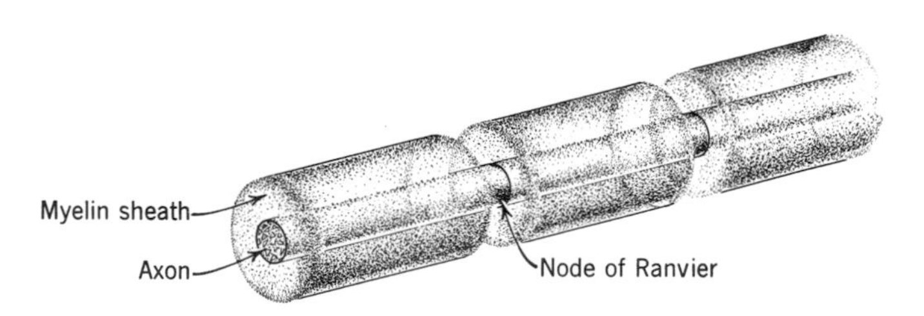

Figure 2-11. Segment of myelinated axon (schematic).

If electrodes were placed within the axon at two adjacent nodes, and if the first node were stimulated to give an action potential, a full-sized action potential would also be recorded from the second node. To see how this happens, suppose that we are able to record at a number of points in the myelinated region between the two nodes, and suppose that the effect of an action potential in the first node is measured at each of these points. Because the internode region has the property of passive spread of potential, at each successive point between the nodes we would record a response much like the one seen at the stimulated node (Figure 2-12), except that the size of the response would be progressively smaller as we move farther from the stimulated node. As far as the myelinated region is concerned, the action potential produced at the stimulated node is no different from any other depolarization and thus spreads passively along the axon in the characteristic decreasing fashion. In fact, the action potential spreads along the internode region into the next node, where, if greater than threshold, it acts just as any other depolarization would, and results in a full sized action potential. Whether the depolarization from the action potential is above threshold or not by the time it has spread passively through the internode region depends of course on how far the nodes are apart; usually, they are close enough together so that an action potential can spread two or three times the average internode distance and still be above threshold. Thus the action potential moves along the axon by passive spread in the myelinated regions and by the occurrence of action potentials at the nodes.

The mechanism by which an action potential is conducted along a nonmyelinated axon is basically the same as that for the myelinated axon, except that it sweeps continuously along instead of "jumping" from node to node: the action potential from one small segment of axon spreads passively to the adjacent segment where it acts as a stimulus to produce another action potential. This same process is repeated for each small length of axon until the action potential has swept over the axon's entire length. The standard analogy used to explain conduction in a nonmyelinated axon is the propagation of a flame along a fire-cracker fuse. The flame in one segment heats adjacent sections of the fuse to

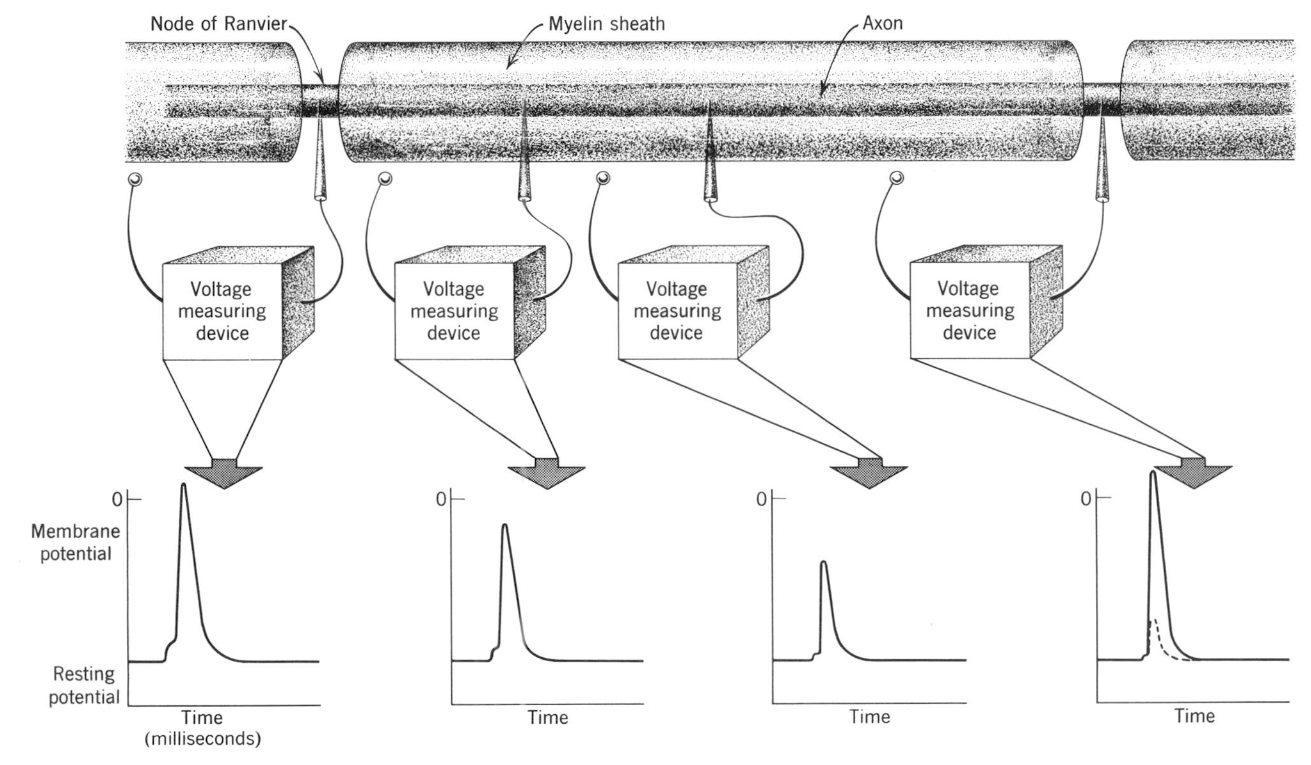

Figure 2-12. Conduction of an action potential along a myelinated axon. The dotted curve in the right-hand graph indicates what voltage would have been recorded at that node if the membrane there had not been able to produce an action potential.

their kindling point, resulting in the spread of flame along the fuse. Just as in the fuse, the action potential is not propagated instantaneously from one end of the axon to the other because it takes a certain small amount of time for the process to spread from one segment to the neighboring region. In most axons, the conduction is quite rapid, however, ranging from 0.1 to 100 meters per second.

As previously discussed, the functioning of the nervous system depends on the efficient transmission of information from one point in the nervous system to another. Passive spread, together with the all-or-none law, provides a means of sending information along the axon: an action potential at one point in the axon spreads passively to neighboring regions where the depolarization acts as a stimulus to produce another action potential, and so on down the axon. The all-or-none law assures us that the action potential will always be full sized, thereby minimizing the chances that it will get lost along the way. On page 20 it was noted that the all-or-none law has a corollary, the existence of a threshold. The fact that there is a threshold is also important since having a threshold which is well above chance variations in membrane potential assures that action potentials will not be spuriously produced.

The all-or-none law, then, is the key property by which rapid and reliable transmission of information is achieved. At the same time, however, the existence of the all-or-none law gives rise to the problem of how information regarding stimulus intensity is to be conveyed. Since graduations in the size of response are not a possibility, some other mechanism for indicating stimulus intensity must be used. An answer lies in the properties of the refractory period and the strength-latency relation which provide for frequency coding: stimulus intensity is coded as number of action potentials moving down the axon per unit of time. Thus the properties of the axon (all-or-none law, threshold, strength-latency relation, and refractory period) together specialize it for transmitting information from place to place within the brain.

Chapter 3

DENDRITES: SPECIALIZATION FOR RECEIVING INFORMATION

In the preceding chapter, a number of properties which specialize the axon for transmission of information were described. In summary we can say that axonal depolarization is transformed into nerve impulses which travel along the axon, and that, in many axons, the frequency of these impulses is an indication of the strength of the depolarizing stimulus. We shall now consider the fate of nerve impulses when they reach the axon's synaptic terminations on a dendrite. The dendrite has certain properties which specialize it for receiving the information transmitted along the axon; often, in fact, at the dendrite, nerve impulse frequency appears to be translated back into a depolarization similar to the one which originally generated the axonal nerve impulses. The problems we must deal with, then, are how the arrival of a nerve impulse at a synapse affects the dendrite, and how dendritic properties can combine to integrate information contained in arriving nerve impulses.

In this chapter, the emphasis will be on developing, in an idealized form, certain basic notions about the operation of synapses. Some of the more important limitations on the accuracy of the concepts presented in this chapter will be discussed at the end of Chapter 4, while the physiological mechanisms underlying synaptic function will be outlined in Chapters 9 and 10.

At the outset, we shall briefly describe the process to which the remainder of the chapter is devoted (Figure 3-1). Neurophysiologists believe that nerve impulses are continually arriving at many synapses, and we shall suppose that, at such a synapse,

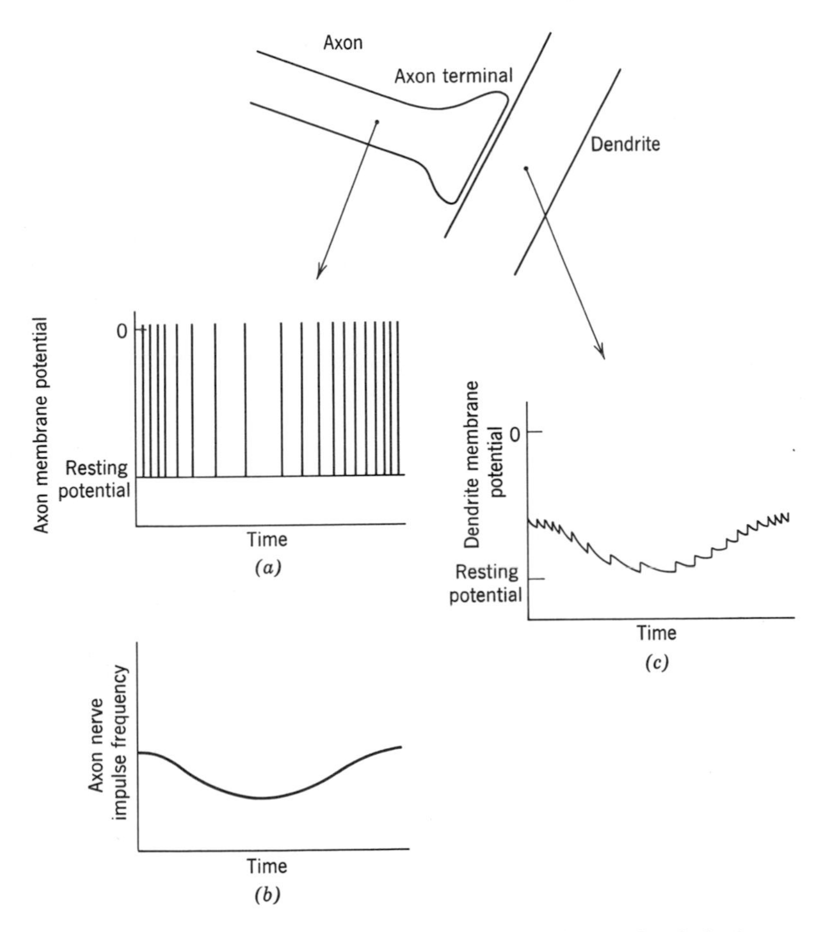

Figure 3-1. Translation of axonal impulse frequency into a depolarization at a dendrite. (*a*) Membrane potential (as a function of time) recorded in axon near terminal. Note that time scale has been so compressed that action potentials appear as vertical lines. (*b*) Nerve impulse frequency (as a function of time) in axon. (*c*) Recording of membrane potential (as a function of time) in the dendrite near the synaptic region. All three graphs have the same time scale.

electrodes have been inserted into the axon near its synaptic termination on a dendrite and into the postsynaptic region of the dendrite itself, in order that membrane potential fluctuations in these regions may be measured. A recording of activity near the axon terminal (Figure 3-1*a*) would typically reveal nerve impulses arriving at a frequency which varied over time (Figure 3-1*b*). Simultaneous records from the postsynaptic electrode would, in the variety of synapses we shall usually deal with, indicate a somewhat jagged variation in the dendritic membrane potential which reflects, at least approximately, the frequency of presynaptic nerve impulse arrivals (Figure 3-1*c*). In other words, there would be a dendritic depolarization whose magnitude is approximately proportional at each instant to the frequency at which nerve impulses are arriving. Thus the frequency of axonal impulses is, at this type of synapse, converted into a maintained dendritic depolarization. An explanation of how this conversion from frequency to depolarization can occur necessitates a more detailed investigation into the effect of a single nerve impulse arrival upon the dendritic membrane potential.

After a nerve impulse arrives at an axon terminal, there is a brief delay of about a half-millisecond, followed by a characteristic fluctuation of the dendritic membrane potential. This fluctuation is known as the **postsynaptic potential** or PSP (Figure 3-2). The typical PSP has a rapid rise to a peak, followed by a much longer (often approximately exponential) return to the resting potential. Particularly striking are two differences between PSPs and action potentials. PSPs are much longer than an action potential, tens or in some cases hundreds of times longer, and also are much smaller in amplitude, again by a factor often as large as several hundred. An understanding of the manner in which PSPs can convert impulse frequency into depolarization requires further investigation into the mechanism by which a PSP is produced, along with a detailed consideration of properties of the dendritic membrane.

The first question which arises, then, is: how does the arrival of a nerve impulse at the synapse produce a PSP? One might conjecture, arguing by analogy to the spread of the action poten-

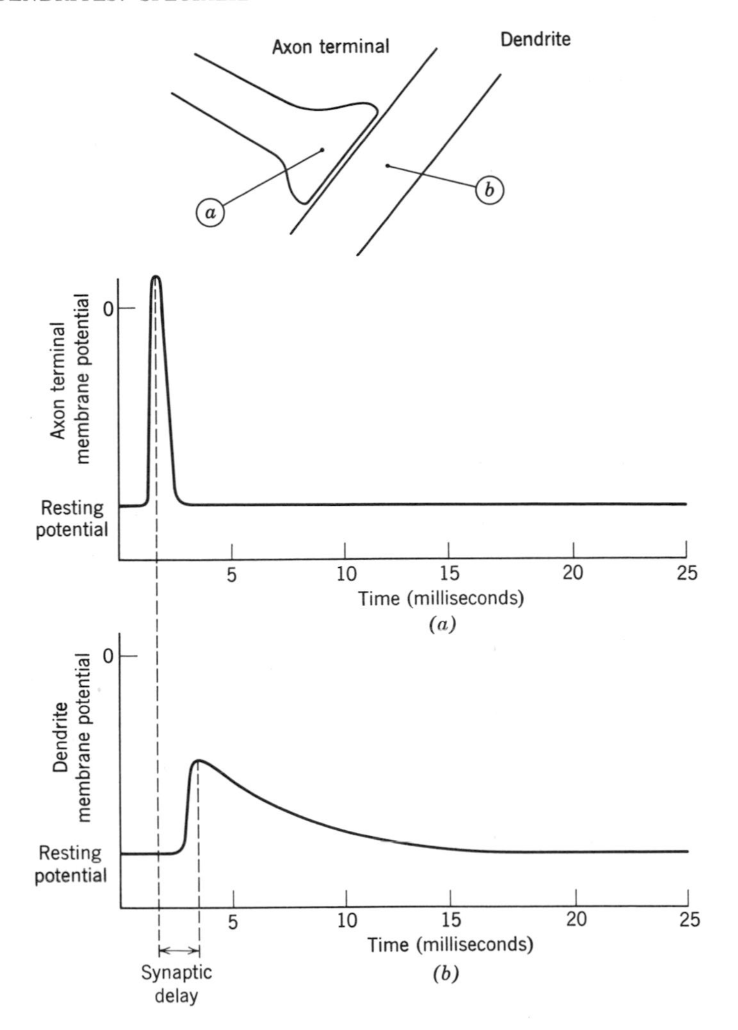

Figure 3-2. Transmission of information at a synapse. (*a*) Recording of membrane potential in axon terminal showing an action potential. (*b*) Recording of membrane potential in dendrite showing a post-synaptic potential.

tial along the axon, that there is an electrical connection between axon and dendrite. This, however, has proved not to be the case.* Rather, it has been shown that the arriving nerve impulse causes

* Although most synaptic transmission involves the mechanisms described in the text, there are a number of special synapses known in which transmission from axon to dendrite is electrical.

the release of a small quantity of a chemical, generically known as a **transmitter substance,** which diffuses across the synaptic cleft and results in the dendritic electrical response. It should be emphasized that this behavior is quite different from that seen in the axon, where application of transmitter does not produce a PSP-like response. The dendritic membrane is thus said to be **chemically excitable** to distinguish this important property from the usual electrical excitability of the axonal membrane. Before further pursuing this matter of chemical excitability, it is necessary to consider the extent to which the electrical properties of the dendrite parallel those of the axon.

Axons and dendrites share a number of electrical properties. The dendrite, like all parts of the nerve cell including the axon, has a resting potential of approximately −60 millivolts; this resting potential is thought to be constant over the length of the dendrite and, indeed, over the entire neuron. In addition, dendrites and axons have the same response to hyperpolarizing stimuli. In both cases, the response mirrors the stimulus (with a slight distortion) and spreads passively away from the site of stimulation in the characteristic manner described in Figure 2-10. It will be recalled from Chapter 2 that the axon's response to small depolarizations is entirely passive, being always a mirror image of the response to hyperpolarizations of similar magnitude. This property, not unexpectedly, is also shared by dendrites. In summary, then, dendrites have all of the *passive* electrical properties of axons.

The distinctive property of the axon is its active response to above-threshold depolarizations, the action potential. For present purposes it may be assumed that dendrites do not respond to depolarization, as do axons, by producing action potentials, but rather show only passive responses to both depolarizing and hyperpolarizing stimuli of all magnitudes.* Thus, for large as well as small depolarizations, the response of a dendrite is the mirror image of the response to hyperpolarizing stimuli. This

* Although the degree to which dendrites respond other than passively to electrical stimuli is still debated, the statement that all are electrically inexcitable is inaccurate. However, it is temporarily assumed, for descriptive purposes, that dendrites do not give action potentials; this matter is more fully discussed on page 63 of Chapter 4.

is illustrated in Figure 3-3 where the responses of a dendrite and axon to various stimuli are compared.

Since the dendrite does not respond to depolarizations, that is, since we have assumed it to be electrically inexcitable, the shape of the PSP must be determined only by the time course of trans-

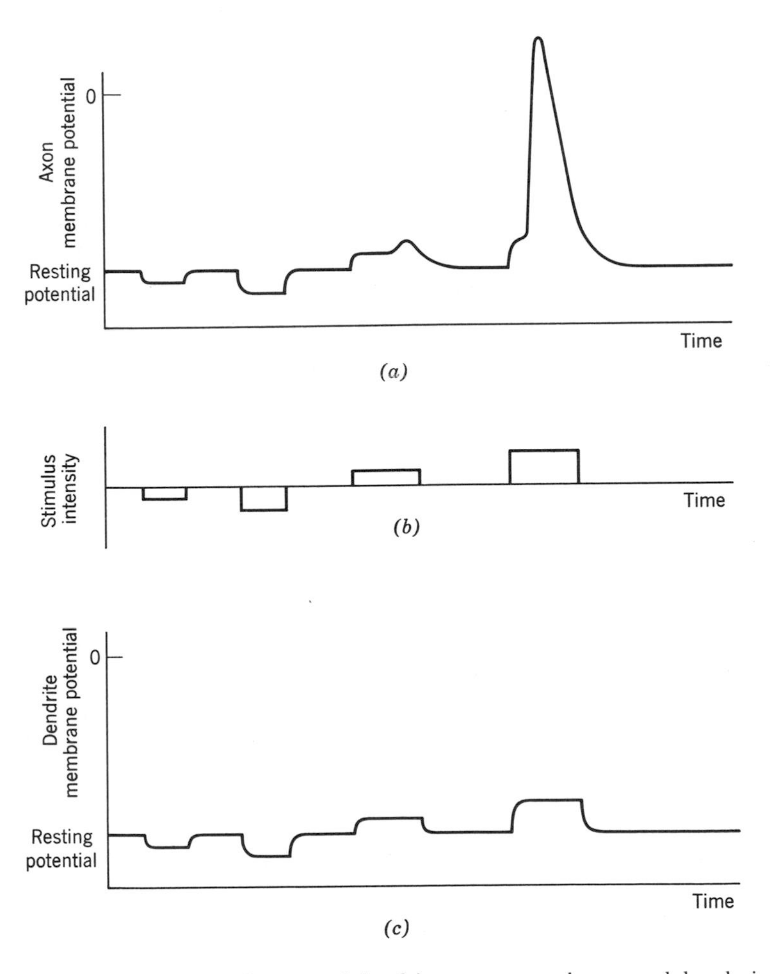

Figure 3-3. Comparison of axon and dendrite responses to hyper- and depolarizations. (*a*) Axon recording showing passive response to hyperpolarizing stimuli, and local response and action potential to depolarizing stimuli. (*b*) Stimulus applied to both axon and dendrite; downward deflection indicates a hyperpolarizing stimulus, and upward deflection a depolarizing stimulus. (*c*) Entirely passive response of dendrite illustrating electrical inexcitability.

mitter concentration in the region of the dendritic membrane.* Although the situation is actually somewhat more complicated (some of the complications will be discussed in Chapters 9 and 10), it may be assumed here that the magnitude of dendritic depolarization is approximately proportional to transmitter concentration. Also, there is an enzyme on or near the postsynaptic membrane which degrades the transmitter at a rate proportional to the amount of transmitter available to be degraded. When a nerve impulse arrives at the synapse, a small amount of transmitter is released, by a mechanism as yet unknown, and diffuses across the 200 or 300 Å synaptic cleft to the dendritic membrane. Since all of the transmitter is released in a very short time, and since the diffusion distance is so short, the entire quantity of transmitter arrives almost simultaneously at the dendrite. Thus the transmitter concentration at the dendritic membrane increases very rapidly to a peak. However, the enzyme which degrades the transmitter also starts at once to remove the active form of the molecule, and the concentration decreases approximately exponentially. (Of course, a certain amount of the decrease of transmitter concentration is caused by diffusion away from the synaptic region, but this process is believed usually to play a secondary role.) Overall, then, a single nerve impulse causes the transmitter concentration to rise rapidly to a peak (after a slight delay for release and diffusion) and then fall more slowly along an exponential time course (Figure 3-4). Since the dendritic depolarization is, for our purposes, approximately proportional to transmitter concentration, it also rises rapidly and returns slowly to the resting potential, thus producing the PSP previously described.

Two important properties of dendrites follow from the fact that dendritic depolarization depends directly upon local transmitter concentration. The first of these may be termed **graded responsiveness,** and the second, **temporal summation.** Graded

* In many neurons, the shape of the PSP is not determined by transmitter properties alone, but instead by the passive electrical properties of the dendritic membrane (see page 131 of Chapter 9 and page 158 of Chapter 10). Also, as indicated by Figure 3-5, depolarization is proportional to transmitter concentration only over a limited range of concentrations. These exceptions, however, are unimportant for the main point of the present argument.

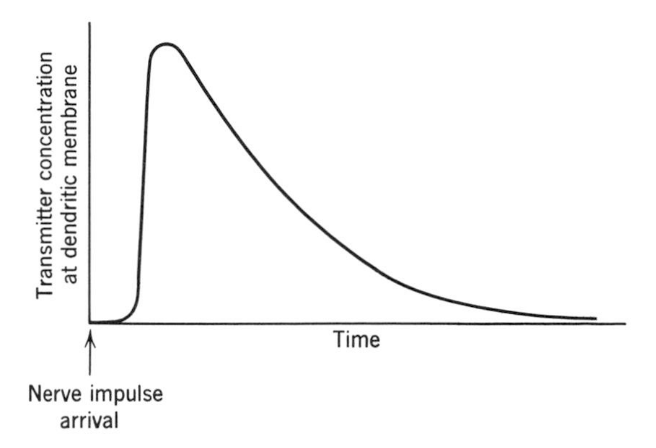

Figure 3-4. Transmitter concentration at a post-synaptic membrane following the arrival of a nerve impulse at the axon terminal. Concentration increases rapidly and then declines in an approximately exponential fashion.

responsiveness refers to the fact that the size of a PSP is proportional to the amount of transmitter released and is to be contrasted with the all-or-none behavior characteristic of the axon. This relationship can be illustrated graphically by plotting peak PSP amplitude as a function of transmitter quantity released (Figure 3-5); the result is a straight line (over an appropriate range of amount of transmitter released) passing through the origin. That this graph has its intercept at the origin points up another difference between the action potential and the PSP: Although the action potential has a threshold, no minimum quantity of transmitter is necessary to initiate a PSP. Thus the notion of

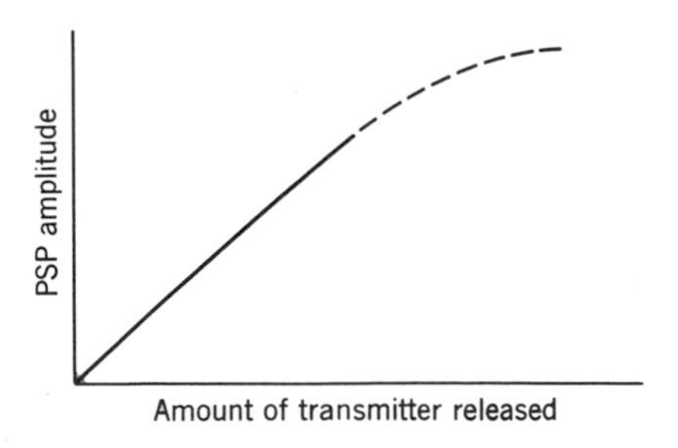

Figure 3-5. Relation between PSP amplitude and quantity of transmitter released by a nerve impulse arriving at the axon terminal.

graded responsiveness may be thought of as contrasting with the concepts of *threshold* and *all-or-none* applicable to axons.

The existence of graded responsiveness implies that the effects of two quantities of transmitter applied simultaneously become added together; from this, it can be inferred that the effects of two quantities of transmitter applied at different times will also be added together. More specifically, one would anticipate that if a nerve impulse arrived at a synapse before the PSP caused by a preceding impulse had completely disappeared, the PSP from the second impulse would simply add to what remained of the first PSP. The same fact may be formulated in a slightly different manner by saying that the dendrite, in contrast to the axon, shows no refractory period. This is illustrated in Figure 3-6 where two PSPs occurring in rapid succession are seen to sum. This important property, the fact that a new PSP simply adds to what remains of all preceding PSPs, is referred to as *temporal summation.*

Earlier in the chapter, it was asserted that nerve impulse frequency can be decoded into a maintained depolarization at the dendrite; we are now in a position to see how, through the properties of the PSP, this can occur. If at a particular synapse, nerve impulses begin to arrive with a constant frequency, each resulting PSP will add to the remainder of all preceding, and an increasing depolarization will occur as the PSPs sum (Figure 3-7). Rather than increasing indefinitely, however, eventually, the depolarization will level off at an average value proportional

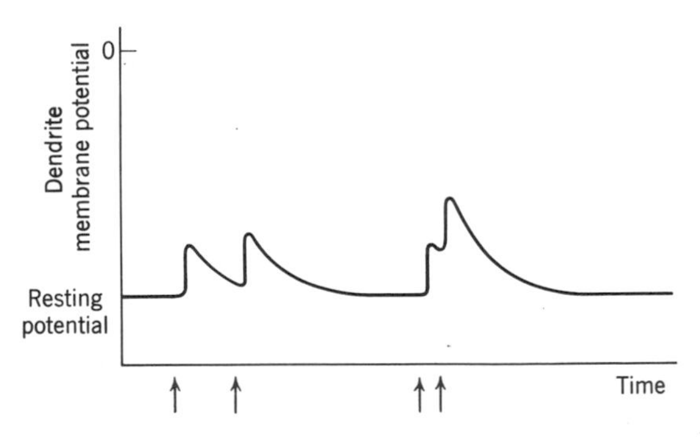

Figure 3-6. Temporal summation of PSPs. Arrows indicate arrival of nerve impulses at the axon terminal.

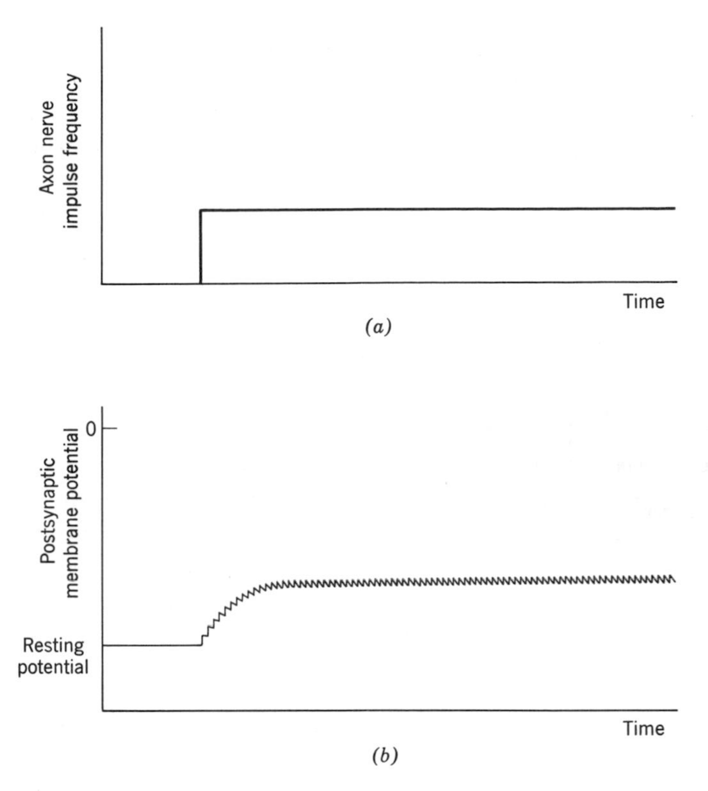

Figure 3-7. Temporal summation of PSPs resulting in a maintained depolarization of the dendrite. (*a*) Frequency of nerve impulses at the axon terminal. (*b*) Membrane potential recorded in the dendrite. Note that the amplitude of the individual PSPs is smaller than in preceding illustrations and that the time scale has been compressed. The amplitude illustrated here is that more commonly seen in the nervous system.

to the frequency at which the nerve impulses are arriving. (This last statement, which will be further justified in a succeeding paragraph, may be temporarily assumed to be correct; later discussion will reveal that it is accurate only over a limited range where linear summation of PSPs holds. See page 64 of Chapter 4.) From Chapter 2 we know that a depolarization in many axons is coded as nerve impulse frequency or, more specifically, that a depolarization results in the generation of nerve impulses at a frequency proportional to the magnitude of the depolarization. Temporal summation, on the other hand, provides the means by which these nerve impulses can be changed back into a depolarization which is again proportional to frequency. Thus the properties of the axon can result in the transformation of depolarization into impulse frequency, while dendritic properties can serve to decode impulse frequency into depolarization again; overall, a depolarization has been moved (perhaps with some distortion) from the axon of one nerve cell to the dendrite of another. How such a process can be used to carry out computations in the nervous system is the subject of the next chapter. For the remainder of this chapter, we shall continue the discussion of temporal summation and introduce some further facts about the PSP.

We must now justify the statement that, when PSPs are arriving at a constant rate, they can sum (linearly) to give an average depolarization proportional to the frequency of arrival. After the PSPs have been arriving for a time, and have reached a steady (average) level, the total depolarization will be the sum of various parts of a large number of PSPs. Suppose the nerve impulses are arriving at the rate of 1000 pulses per second; thus, there is a 1-millisecond interval between arrivals. Furthermore, imagine a PSP divided up into successive parts by 1-millisecond intervals (Figure 3-8*a*). Consider the 1-millisecond interval starting with the beginning of some particular PSP; we wish, for reasons which will soon become apparent, to calculate the total area under the curve relating dendritic depolarization to time for this particular 1-millisecond interval (shaded area in Figure 3-8*a*). One way of calculating this area is immediately evident: the area is given by the product of the mean depolarization for the interval and the length of the interval, that is average depolarization *times* 1 millisecond *equals* area under the depolarization-time curve for the 1-millisecond time interval under consideration.

A second and somewhat more complicated calculation of the same area leads to the desired relation between average depolarization and frequency. During the 1 millisecond under consideration, the area is made up of the sum of the areas contributed by the remainder of all PSPs persisting from preceding times and by the first millisecond of the PSP which started at the beginning of the interval; stated in a slightly different way, the

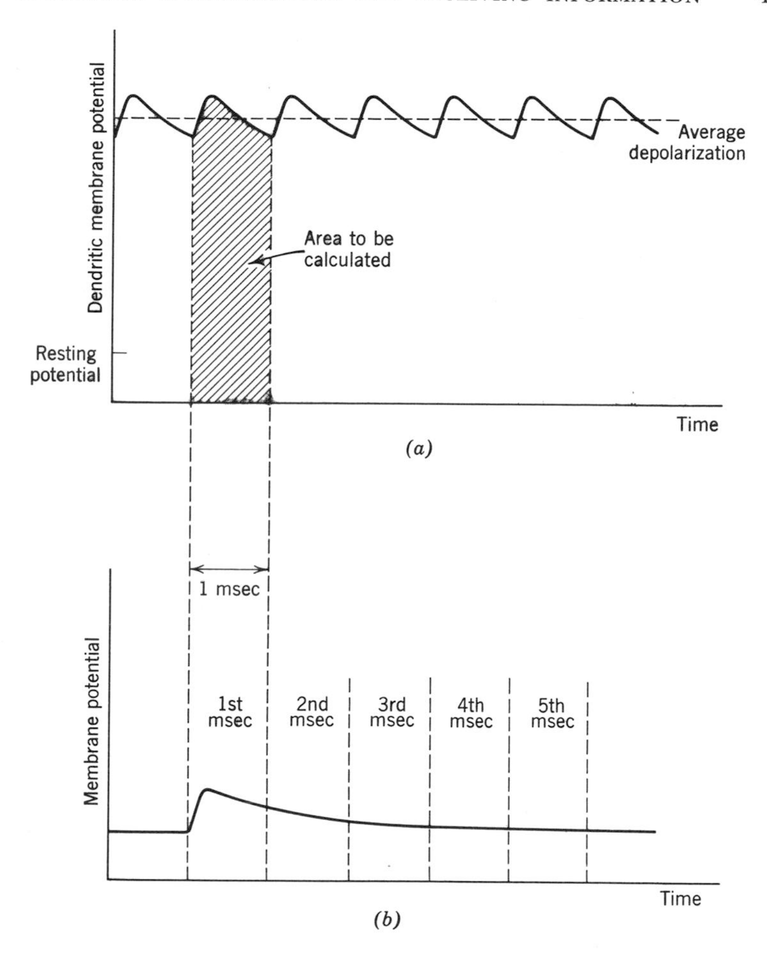

Figure 3-8. Detail of maintained depolarization produced by temporal summation of PSPs. (*a*) Dendritic membrane potential with area to be calculated; PSPs occur here at a rate of one per millisecond. (*b*) Single PSP divided into one millisecond segments.

area will be the sum of the areas contributed by a 1-millisecond segment of a large number of PSPs. The PSP starting at the beginning of the interval will contribute the area of its first millisecond. The PSP starting at the beginning of the preceding interval will, since it started a millisecond earlier, contribute the area of its second millisecond. In general, the PSP starting in the *n*th preceding interval will contribute the area of its *n*th millisecond. Altogether, since the temporal summation has been assumed to have gone on for a long time (compared to the duration of a single PSP), the area under the 1-millisecond interval of the curve relating depolarization to time will equal exactly the total area under a single PSP. We have thus calculated the same area in two different ways, and so average depolarization *times* one millisecond *equals* the area under a single PSP. Dividing both sides of the equality by one millisecond, and remembering that the frequency of arrivals is 1000 per second (which equals the reciprocal of one millisecond), we have the equation average depolarization *equals* frequency *times* the area of a single PSP. Average depolarization is thus proportional to frequency, with the proportionality constant being the area of a PSP.

To this point, PSPs have been described as depolarizing the cell. Neurons, however, characteristically display a second type of PSP which is hyperpolarizing rather than depolarizing. The hyperpolarizing variety has a shape much like that of the depolarizing PSP, except that it is inverted; that is, it (except in the situation described on page 64) consists of a rapid hyperpolarization followed by a relatively prolonged return to the resting potential (Figure 3-9). Since depolarization of the neuron results in the production of nerve impulses, the depolarizing PSP is termed an **excitatory postsynaptic potential** usually

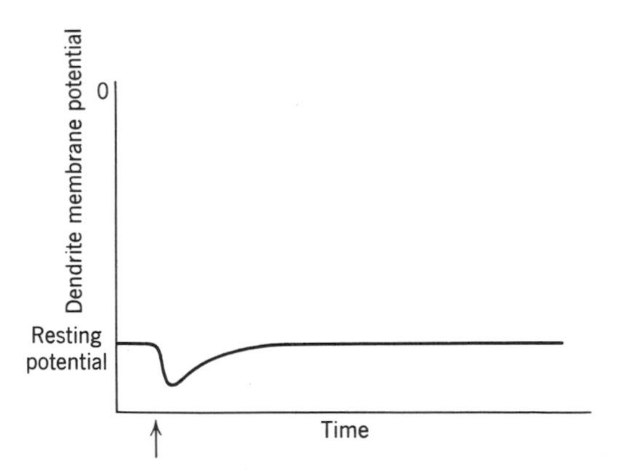

Figure 3-9. An inhibitory post-synaptic potential. Nerve impulse arrived at the axon terminal at the time indicated by the arrow.

abbreviated **EPSP.** The hyperpolarizing PSP, on the other hand, is known as an **inhibitory postsynaptic potential** (or **IPSP**) because it opposes the action of EPSPs and tends to prevent the generation of action potentials.

Since there are two varieties of PSPs, the question arises at once whether there are also two varieties of action potentials, one giving rise to EPSPs and the other to IPSPs. There is, as it happens, only a single type of action potential, the type described in Chapter 2. Whether an action potential gives rise to an EPSP or IPSP depends entirely upon the properties of the synapse; that is, some synapses are specialized to produce EPSPs and others, IPSPs. Although the evidence is not entirely clear as yet, it appears that a given axon liberates at all of its synaptic terminals only one particular type of transmitter substance,* and that whether this transmitter produces an EPSP or an IPSP depends upon the characteristics of the postsynaptic membrane. However, it is accepted that a synapse must be of one or the other variety and cannot be inhibitory at some times and excitatory at others.

That EPSPs have properties which permit temporal summation is of central importance in the operation of the nervous system. It is therefore of interest to know how the properties of IPSPs compare with those of the EPSP; in particular, it is important to know if IPSPs exhibit temporal summation. Except for the fact that one is depolarizing and the other is hyperpolarizing, EPSPs and IPSPs do indeed share the same properties. Thus IPSPs show temporal summation, and, if they occur at a constant rate, IPSPs also can sum to give an average *hyper*polarization proportional to the frequency of nerve impulses causing them. Whereas EPSPs can transform nerve impulse frequency into a maintained depolarization, IPSPs can convert nerve impulse frequency into a maintained *hyper*polarization.

To summarize: while dendrites and axons have the same passive electrical properties, the active response of the postsynaptic portion of the dendritic membrane differs considerably from that

* Numerous transmitter substances (one of them being acetylcholine) have been tentatively or quite definitely identified. As yet, however, the chemistry of their action is very poorly understood. The reader interested in pharmacological aspects of synaptic transmission should consult the monograph by McLennan.

of the axonal membrane. A nerve impulse arriving at a synapse results in a PSP in the dendrite, and the key property of this PSP is that it is a graded response; since it is graded, temporal summation is possible, and nerve impulse frequency may be translated into either a depolarization or a hyperpolarization, depending upon the nature of the synapse. So far, then, we have seen how the axon is specialized (by converting depolarization into impulse frequency, for example) to transmit information, and how the dendrite can act (through the transformation of impulse frequency into a de- or hyperpolarization) to receive the information from the axons. How these processes can form a basis for computation within the nervous system is the subject of the following chapter.

Chapter 4

COMBINATIONS OF EXCITATION AND INHIBITION: COMPUTATION IN THE NERVOUS SYSTEM

Neurophysiologists have compiled a considerable store of information on the properties of neurons, but much less is known about how these properties are used to carry out the complicated neural computations underlying the organism's behavior. In this chapter we shall present one theory of how the basic computations in the nervous system are performed, in order to place the preceding description of neuronal structure and properties in a broader context. Although evidence is good that something very close to what we shall describe is indeed going on in the vertebrate brain, the precise limits of accuracy of what we shall say and how far our theory goes in explaining nervous system function are still under investigation. At the end of the chapter we shall briefly discuss the range of applicability of the theory and indicate some of the directions a more comprehensive treatment of neural integration must take.

Certainly a very striking feature of the neuron is the large proportion of its membrane area receiving synaptic contacts from other nerve cells. To the neurophysiologist this anatomical fact means that a neuron may receive information simultaneously from a large number of other neurons. At the same time, the nerve cell body characteristically gives rise to only a single axon and thus to a single channel through which information flows out of the neuron. (This axon typically branches, of course, but the same information flows over all the branches.) A neuron, then, receives information from a number of sources and must integrate this information to determine what emerges from the single output. In the preceding chapters many of the steps in this process of integration and information transfer have already been discussed. We have seen, for example, that the nerve impulse frequencies at the various excitatory and inhibitory synapses of a neuron can be converted into depolarizations or hyperpolarizations and also that the axon can reconvert depolarization into impulse frequency for transmission to still other neurons. Remaining, however, is the question of how depolarizations and hyperpolarizations arising at the various synapses of a neuron can combine to produce some net potential at the origin of the axon where nerve impulses are generated. In the remainder of this chapter, we shall concentrate upon the question of how information from many sources can be combined and also upon an overall process of integration in the nervous system.

A preliminary understanding of the integration process we shall describe depends upon knowing only one additional property of neurons: EPSPs and IPSPs arriving over different synapses sum, to a good first approximation, simply algebraically. This property, known as **spatial summation,** should not be surprising in view of the existence of temporal summation; since PSPs from a single synapse add together it is entirely reasonable to expect the same behavior of PSPs from different synapses. Thus various inputs to a neuron combine by summation to determine the single output; how this, together with previously described properties, can be used as a basis for computation in the nervous system is the central point of the following discussion.

Before proceeding to a discussion of the integration process, however, it is necessary to explain a point which has been implied in the preceding chapters. A sharp distinction has been drawn between the properties of dendrites and axons by stating that dendrites are electrically inexcitable and do not give rise to action potentials. Later, we must return to this question, for the distinction has been made oversharp; but for the time being we wish to emphasize this fundamental difference in dendrites and axonal properties and consider one of its implications. If the dendrites are inexcitable while the axons give rise to nerve impulses, there must be some region of the cell closest to the dendrites which is electrically excitable. This region is believed to be at the **axon hillock,** the junction between soma and axon. Thus a depolarization arising in a dendrite spreads passively down the dendrite, through the soma and finally to the axon hillock where nerve impulses can be generated. The axon hillock, then, is the point in a neuron where all nerve impulses arise. That this fact is important to the operation of the nervous system will become apparent after the following discussion.

THE SLOW POTENTIAL THEORY

For the sake of clarity we shall describe neural integration according to the slow potential theory in several specific hypothetical situations. Suppose, first, that a neuron—designated *E*—sends its axon to form a synapse with the dendrite of another neuron *N* (Figure 4-1), and suppose further that the *E-N* synapse is excitatory (depolarizing). We shall imagine that there is in the soma of *E* an above-threshold depolarization whose magnitude changes relatively slowly; such a deviation from the resting potential which varies relatively gradually is known as a **slow potential**—*slow* because changes in membrane potential are much less rapid than those seen in the action potential. For the present, we shall not worry about how such a slow potential normally arises. (It may, for example, be temporarily supposed that the slow potential is the result of a stimulus applied by the experimenter.) From our discussion of axons we know that the slow potential in *E* will be converted into a nerve impulse frequency in *E*'s axon and that the nerve impulses will travel rapidly to the axon terminations in the *E-N* synapses. At the dendrite of *N* the nerve impulse frequency will be translated into a slow potential by temporal summation of EPSPs. This slow potential will have

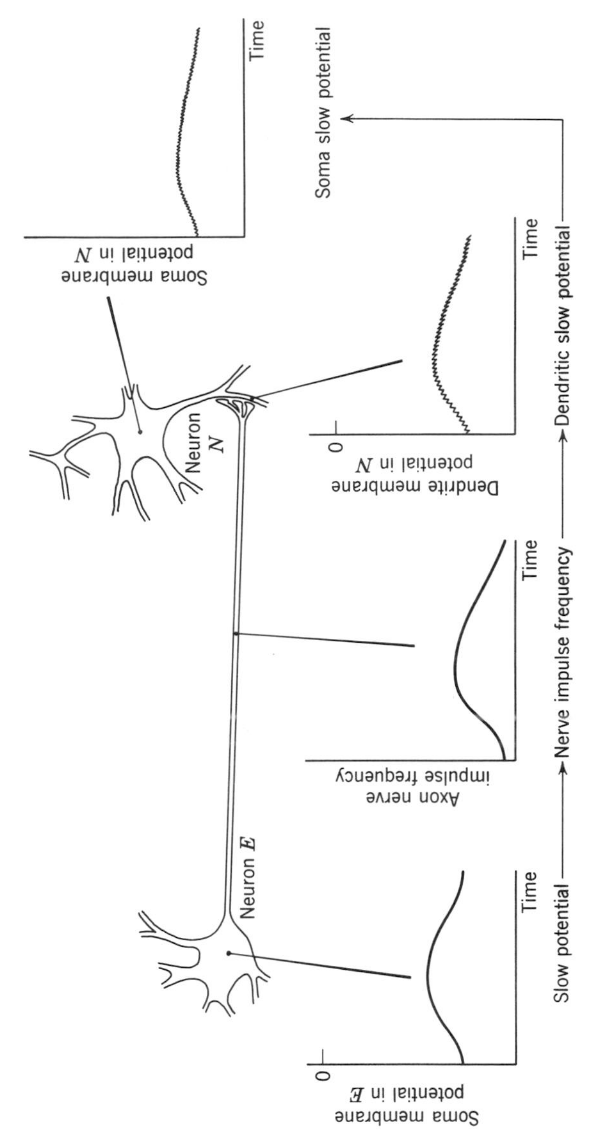

Figure 4-1. Information transfer according to the slow potential theory. A slow potential in the soma of neuron *E* is coded as nerve impulse frequency and then decoded into a slow potential again in the dendrite of neuron *N*. This dendritic slow potential spreads passively to the soma of *N*.

a shape much like the slow potential in *E* which was originally coded as impulse frequency, although the amplitude may be quite different. The end result, then, is that the slow potential in the soma of one neuron (*E*) has been transmitted more or less intact to the dendrite of another neuron (*N*).

Since dendrites show all the passive electrical properties of axons, they, of course, exhibit the phenomenon of passive spread of potential. Thus the slow potential in *N*'s dendrite will spread passively down the dendrite to the soma and from there to the axon hillock region of *N*'s axon. A phenomenon seen in passive spread is that the size of the slow potential decreases with distance from its origin; however, if the slow potential spreading from the dendrite of *N* is still of sufficient magnitude (that is, above threshold) by the time it reaches the axon hillock region, it will result in the generation of nerve impulses and will be coded as impulse frequency to be sent along to still another neuron.

Through a series of steps similar to those just described, it would be possible to transmit a slow potential from neuron to neuron throughout the entire nervous system. This particular case, however, is neither very interesting nor very realistic. In the normal operation of the nervous system, large numbers of axons from various different nerve cells converge upon a single neuron which must integrate the slow potentials arriving from each of these sources. A more realistic situation to consider, and one which better reveals the nature of the integration process we are describing, is that of two different neurons, *E*1 and *E*2, making excitatory synapses with different dendrites of the neuron *N* (Figure 4-2). Suppose that *E*1 and *E*2 have different slow potentials in their somas. These slow potentials will be coded as impulse frequency and transmitted to the dendrites of *N*, where each will be decoded into a replica of the original slow potential. As in the preceding example, each slow potential will spread passively down its dendrite to the soma. In the soma the two slow potentials will simply sum, and it will be this sum which spreads to the axon hillock where it may be coded as nerve impulse frequency for transmission to still another neuron. In summary, then, slow potentials in two neurons are transmitted to a third neuron where

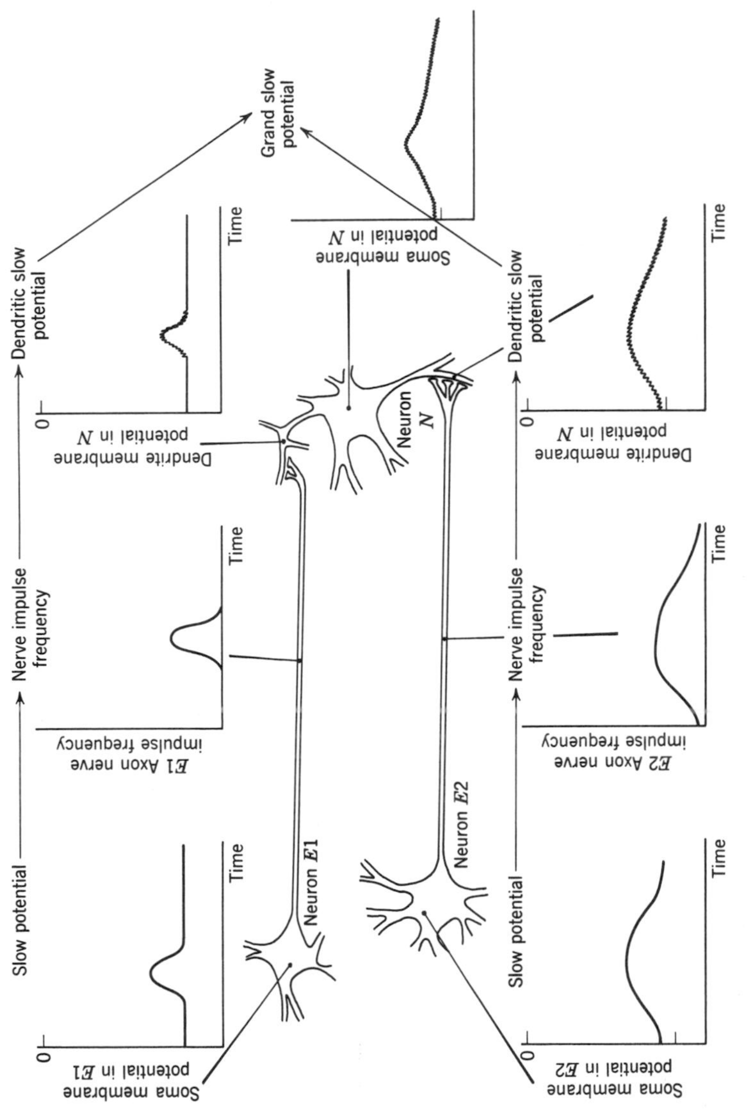

Figure 4-2. Integration of information from two neurons: forming the sum of two slow potentials in neuron *N*.

they are added together. The nervous system thus has a mechanism for summing slow potentials.

In this example, both synapses were assumed to be excitatory. We shall now consider the situation in which one is excitatory and the other is inhibitory. Let *E* be the neuron making an excitatory synapse with *N*, and *I* the neuron with an inhibitory synapse (Figure 4-3). Furthermore, suppose that *E* and *I* have different slow potentials determining their production of nerve impulses. The slow potential in *E* will be transferred to the dendrite of *N* just as in the preceding examples, and will spread passively from there to *N*'s soma. *I*'s slow potential will also be transferred to *N*'s dendrite but will be inverted; that is, the slow potential in *N*'s dendrite will have approximately the same shape as the original slow potential in *I*, but it will be in the hyperpolarizing rather than the depolarizing direction since it is reconstructed by the temporal summation of IPSPs. This inhibitory slow potential, like the excitatory one, spreads passively to the soma of *N*, where it will sum with the depolarizing slow potential transmitted from *E*. Since one of these two slow potentials is depolarizing and the other is hyperpolarizing, the inhibitory slow potential will subtract from the excitatory one, and the difference between the two will be coded as nerve impulse frequency for transmission to another neuron. Overall, then, slow potentials in two different neurons have been transferred to a third neuron where one has been subtracted from the other. The nervous system thus has a mechanism for subtracting one slow potential from a second.

It is not too difficult to see from these examples what will happen in still more complex cases. Slow potentials transmitted to the same dendrite will simply sum in the dendrite, and this sum will spread passively to the soma, where it in turn will sum with slow potentials spreading from other dendrites. In the soma, the grand sum of all slow potentials from all sources will be coded as nerve impulse frequency to be transmitted to still other neurons.

Up to this point we have described a number of properties of neurons, and have seen how they can be used to perform the simple but basic operations of negation and summation. These

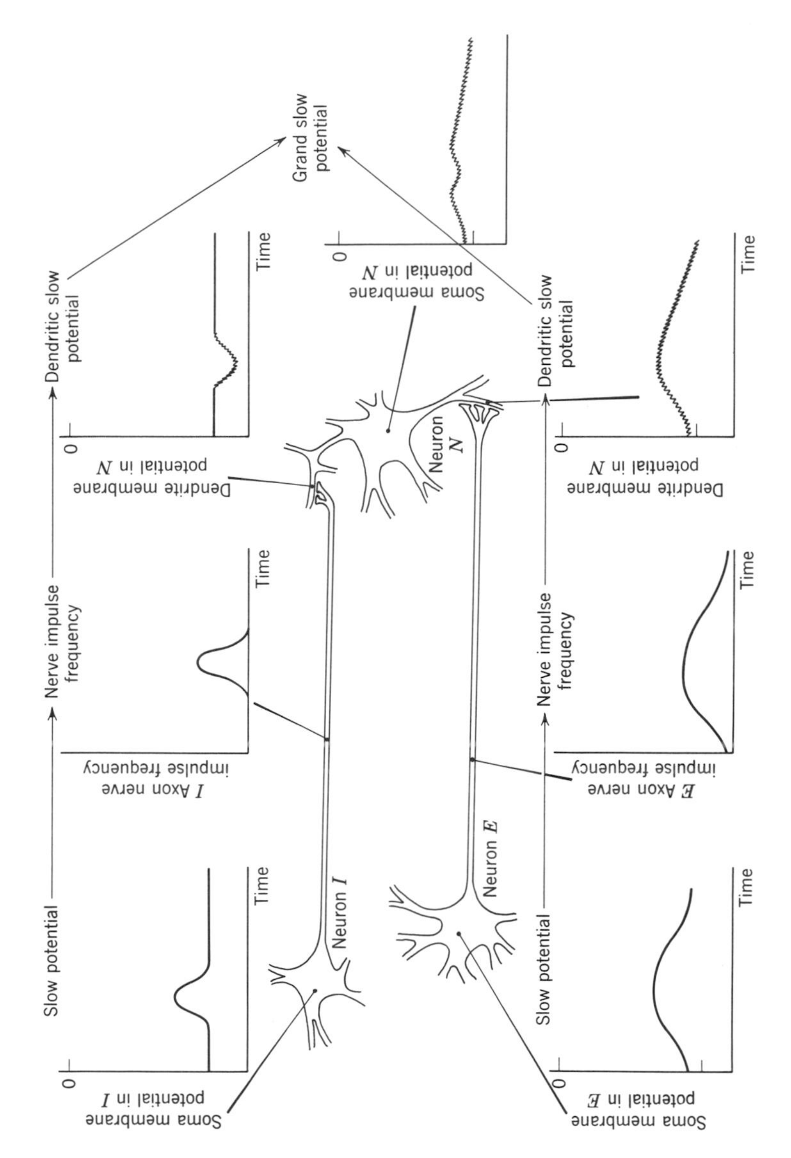

Figure 4-3. Integration of information from two neurons: forming the difference between two slow potentials in neuron *N*.

neuronal properties, taken together, underlie a chain of steps (slow potential converted to impulse frequency, and frequency back to slow potential) which have as their result the summation, in one neuron, of different slow potentials transferred from various other neurons. This, in broad outline, may be taken as a description of principles fairly generally accepted by neurophysiologists as a possible basis for computation in the nervous system. Having presented the essentials of this integration process, we shall first consider certain limitations and extensions of the main idea and then discuss some idealizations that have been made previously.

LIMITATIONS AND EXTENSIONS OF THE SLOW POTENTIAL THEORY

Because neurons have a threshold, only the above-threshold part of a slow potential is coded as nerve impulses to be sent on to another neuron. This means that if the sum of slow potentials arriving at a particular neuron happens to be less than threshold, the information contained in the sub-threshold slow potential will be simply lost rather than passed to other neurons for further processing. Furthermore, the threshold introduces a striking asymmetry in the response of a neuron to sums of EPSPs as opposed to IPSPs. A sufficiently large sum of EPSPs can be coded as nerve impulse frequency for transfer to another neuron, whereas IPSPs, regardless of magnitude, never produce nerve impulses. However, it becomes quite clear that if the neuron is to act on all information arriving at its synapses, both depolarizing and hyperpolarizing slow potentials must have an effect on the output of the nerve cell. In the usual operation of the nervous system this is thought to be achieved by "spontaneous activity" of the neuron. That is, the anatomical and physiological arrangement of nerve cells appears to be such that there is typically a sufficient number of EPSPs arriving to produce a steady, above-threshold depolarization. All inputs to the cell, then, serve to modify this baseline depolarization. Although an inhibitory slow potential will diminish the depolarization, the net slow potential in many neurons will not generally fall below the threshold for nerve impulse generation. In this way, information about both hyperpolarizing and depolarizing slow potentials can be transmitted to other neurons and thus is not lost.

It has already been pointed out (page 51) that when a slow potential is transferred from one neuron to another, its shape (relative magnitude as a function of time) is preserved,* but its absolute amplitude is not. Since changes in amplitude can be essential to the proper functioning of the nervous system, we now consider four important factors which determine the amplitude alterations suffered by a slow potential when it is transferred from one neuron to the next.

As was stated in Chapter 3, the average slow potential magnitude at the postsynaptic side of a synapse is, under certain circumstances, proportional to the frequency at which nerve impulses are arriving at the axon terminal; furthermore, the proportionality constant is equal to the area of a single PSP (page 44). PSP area, in turn, depends upon the amount of transmitter released per nerve impulse (see Figure 3-5): the larger the quantity of transmitter released, the larger the area of the PSP. Since larger synapses presumably secrete larger quantities of transmitter, one of the important factors determining the size of a slow potential transmitted to a neuron is the size of the synapse doing the transmitting.

It is typical for an axon to divide into smaller axons before terminating on a neuron, so that one neuron may contribute a number of synapses to a second neuron (see page 1). Since each of these synapses has nerve impulses arriving at the same frequency, each transmits the same slow potential to the second neuron. The slow potentials from all of the synapses sum, so that the final amplitude of the slow potential transmitted to the second neuron is directly related to the number of synapses contributed to that neuron by the first neuron.

Synapses are distributed over the surface of a neuron's soma and

* In practice, the shape of a slow potential is perfectly preserved only under certain restricted conditions, some of which are indicated in the next section. For descriptive purposes, however, we shall temporarily assume that a slow potential is transmitted with no distortion other than change in amplitude.

dendrites (see page 3). Because dendrites may be relatively long, perhaps a millimeter in some cases, some synapses are quite remote from the soma, others are comparatively near, and still others are directly on the soma. As discussed earlier in the chapter, the information finally transmitted from one neuron to another relates to slow potential which reaches the nerve impulse-generating region at the origin of the axon. This means that a slow potential arriving from a synapse far out on a dendrite will be quite attenuated at the soma (it has spread passively over a relatively large distance; see Figure 2-10*b*), whereas a slow potential arriving from a synapse on the soma will not be attenuated. In other words, distant synapses contribute smaller slow potentials than otherwise similar synapses on or near the soma.

Thus far we have considered factors that influence the amplitude of a slow potential decoded from a given impulse frequency. However, the amplitude changes suffered by a slow potential in transmission from one neuron to another depend not only on how the nerve impulse frequency is decoded into a slow potential, but also on how the slow potential in the first neuron was translated into impulse frequency. As pointed out on page 23, a certain depolarization in one neuron may result in a ten-per-second nerve impulse frequency, whereas the same depolarization in another neuron would cause it to produce nerve impulses at 500 per second. Clearly, other things being equal, the lower frequency neuron will transmit slow potentials which, when decoded, will be much smaller than those transmitted by the higher frequency neuron.

Altogether, then, there are four principal factors governing the amplitude of a slow potential in one neuron as compared to its amplitude in the neuron from where it came. If a (1) high-frequency neuron contributes (2) many (3) large synapses which are (4) close to the soma of a second neuron, the transmitted slow potential will in general have a large amplitude. On the other hand, if a (1) low-frequency neuron contributes (2) few (3) small (4) distant synapses, a low amplitude slow potential will result. Any combination of these factors may, of course, operate in a particular situation to determine slow potential amplitude.

In the usual operation of a typical neuron, many different slow potentials arriving from other neurons are all summed together to form a grand slow potential. This grand slow potential, in turn, is coded as nerve impulse frequency for transmission to still other nerve cells. When the grand slow potential is formed from the constituent slow potentials, the weight given to each constituent in the final sum is determined by the factors enumerated above. Because these factors are consequences of anatomical or physiological properties of the neurons involved, they are more or less fixed; that is, in order to change the weight given to one particular constituent slow potential, it would be necessary, for example, to grow more synapses. It should be clear that considerably more flexibility in nervous system function could be achieved if the weights given to a particular slow potential were under neural control. Such a process, known under the name of **presynaptic inhibition,** does in fact occur in the nervous system. The mechanism of presynaptic inhibition is based upon a property of axon terminals illustrated in the following type of experiment.

Experiments in which one electrode has been inserted into an axon terminal and another into the dendrite near the postsynaptic membrane have revealed that the amount of transmitter released from the terminal per nerve impulse depends upon the value of the axon terminal potential between nerve impulses. More specifically, if the axon terminal is hyperpolarized slightly for a period of time, each nerve impulse arriving during that period causes a PSP slightly larger than usual, whereas if the terminal is slightly depolarized, each nerve impulse causes a PSP smaller than normal. Figure 4-4 presents the effect of steady membrane de- or hyperpolarizations upon the amount of transmitter released by the superimposed action potentials. This graph is valid, of course, only for a range of depolarizations below threshold for the production of nerve impulses on the one hand, and for hyperpolarizations insufficient to prevent action potentials on the other. It must be emphasized also that the graph in Figure 4-4 is applicable only to very slowly varying de- or hyperpolarizations, ones in which the membrane potential is essentially constant during the period between action potentials.

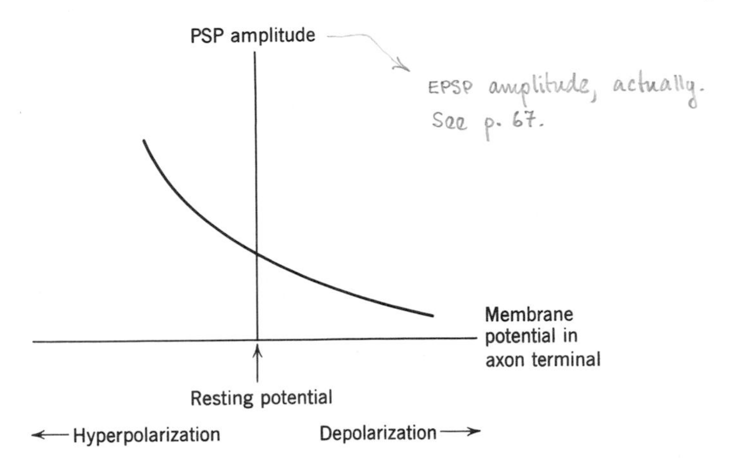

Figure 4-4. Effect of axon terminal membrane potential on PSP amplitude. The ordinate gives peak PSP amplitude, and the abscissa gives the axon terminal membrane potential between action potentials.

By displacing the axon terminal membrane potential slightly from its resting value, it is possible, then, to alter the size of PSPs transmitted by a synapse, and thus to control the overall amplitude of its slow potentials. The nervous system has exploited the possibilities inherent in this fact by using it to modify the weight given to the PSPs transmitted by a particular synapse. By transmitting a slow potential to the axon terminal (via an axo-axonal synapse), the amount of transmitter secreted by that terminal (and therefore the resultant PSP amplitude) is altered depending on the slow potential magnitude in the axon terminal. If, for example, an axon terminal is depolarized (by the temporal summation of EPSPs transmitted over the axo-axonal synapse), the PSP transmitted by that axon terminal would be attenuated.

To date only depolarizing axo-axonal synapses have been discovered so that slow potentials transmitted by these axon terminals may only be decreased from their normal amplitude. For this reason, the mechanism for modifying the weight given to synapse's slow potential has been termed *presynaptic inhibition.* (*Inhibition* because the mechanism decreases the amplitude of the synapse's PSP, and *presynaptic* because the PSP amplitude is decreased by events occurring in the axon terminal rather than by a postsynaptic event as is the case with ordinary inhibition.)

It is instructive to compare the properties of presynaptic inhibition and ordinary (or postsynaptic) inhibition. Postsynaptic inhibition works by subtracting a PSP from whatever other PSPs happen to be arriving at the neuron. Thus, it is nonselective (in that it affects all of the neuron's slow potentials equally) and subtractive (because it decreases other slow potentials by subtracting from them). Presynaptic inhibition, on the other hand, is selective (affecting only the PSPs arriving over particular synapses) and multiplicative (decreasing a particular slow potential to, for example, one-half of its normal uninhibited amplitude regardless of what that amplitude is). Without going into details, it is interesting to note in passing that presynaptic inhibition (or a similar mechanism) provides a means by which the nervous system can multiply two slow potentials rather than simply adding or subtracting them as we have described previously.

A discussion of the weights assigned to the various components of a neuron slow potential leads to a consideration of the special case in which there is a single dominant component with so large a weight that the arrival of only a single impulse results in an above-threshold EPSP. This might occur, for example, if there were a large synapse very near the impulse-generating region of the neuron or if an incoming axon branched so many times that its terminations formed a large proportion of the neuron's synapses. In either case, the arrival of a single nerve impulse over the axon (and, of course, its branches) would give rise to one or more outgoing impulses. It must be emphasized that since there is a relatively direct coupling between input and output, such behavior is really quite different from that described previously. Synapses which operate in this one-to-one fashion are indeed known, but present evidence indicates that the majority of synapses within the nervous system probably are not of this variety. It is also important to note that the two types of behavior are not mutually exclusive; it is possible that a fraction of the synapses on a neuron act to decode impulses into slow potential while others are of the one-to-one variety. By considering all of the gradations between the two extremes (one-to-one synapses ranging to "slow potential" synapses) we find a large variety of behaviors of nerve cells. To what extent these different varieties occur in the nervous system is not yet known.

DISCUSSION OF SOME IDEALIZATIONS

By now we should have some insight into the richness of the neuronal properties used by the nervous system in its computations. Much of this information has, of course, been presented in a quite schematic and simplified form. In addition to the limitations and extensions of the theory just discussed, we should consider certain idealizations made throughout the presentation. It has been tacitly or explicitly assumed that, (1) the frequency at which a neuron produces action potentials depends only on membrane potential and not at all on the length of time that particular value of membrane potential has obtained, (2) dendrites and cell bodies do not produce nerve impulses, and (3) PSPs sum linearly. None of these assumptions are strictly true and for some situations not even approximately true.

In some neurons the first assumption does appear to hold true, but perhaps more commonly a neuron initiates nerve impulses at a higher frequency in response to an increasing stimulus than in response to a nonchanging one of comparable magnitude. In fact, neurons may be classified into at least two separate categories on the basis of their response to steady or changing stimuli. Neurons of the first type, **tonic neurons,** have been dealt with almost exclusively so far. These cells will produce impulses in response to a constant stimulus and are to be distinguished from neurons of the second type, **phasic neurons,** which respond only to changing stimuli and not at all, or at least only briefly, to a constant stimulus. It is tempting to conclude that the phasic neurons serve to add differentiation to those more elementary operations discussed earlier in the chapter. However, the detailed understanding of these neurons is as yet insufficient to warrant assigning them a specific role in nervous system function.

The preceding paragraph has indicated that it is typical for a neuron to discharge impulses at a higher rate in response to an increasing depolarization than to a nonchanging depolarization of the same magnitude. This is obviously the case for the phasic neurons which have a zero frequency if the depolarization is constant, but is typically true for many tonic neurons as well (Figure 4-5). It appears that impulse frequency in a number of tonic neurons is jointly proportional to depolarization and rate of change of depolarization, at least approximately. Without discussing the matter in detail, it should be pointed out that, because of the shape of the individual PSP, the requirement that a slow potential be coded as impulse frequency and then decoded by temporal summation *without change in shape* can be shown to imply that the frequency should be proportional to the sum of the depolarization and its first derivative. Thus the behavior of tonic neurons is qualitatively in agreement with what would be expected if slow potentials were transferred from one neuron to the next without

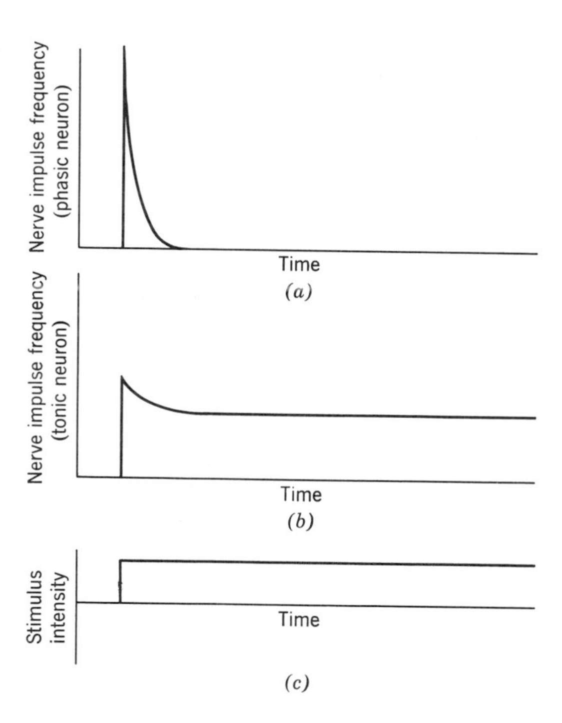

Figure 4-5. Nerve impulse frequency in response to a steady stimulus. (*a*) A phasic neuron. (*b*) A tonic neuron. (*c*) Stimulus intensity as a function of time which produced the responses illustrated in (*a*) and (*b*).

distortion. However, the quantitative details of this process still need further examination.

In the view of nervous system function outlined here, the inability of dendrites to produce an action potential has played an important role. If dendrites did give action potentials in the same manner as axons, the output of the neuron would not be determined by the sum of all incoming PSPs, but rather nerve impulses would arise independently from various dendrites whenever the PSPs exceeded threshold. In such a situation the integrative activities of the neuron, particularly where inhibitory PSPs are concerned, would be circumvented. However, dendrites may support action potentials and, at the same time, be operative for the sort of integrative scheme described in the first part of this chapter; this can occur if the threshold of the dendrites is considerably higher than it is for the axon hillock region. According to this alternative, the sum of all PSPs would determine the neuron's impulse frequency, but the peak of each impulse originating at the axon hillock might also be above threshold for the dendrites, thereby causing a dendrite action potential. Thus it would not be required that dendrites be inexcitable, but only that they be sufficiently less excitable than the axon hillock.

In practice, at least a portion of a typical neuron's dendritic tree apparently does produce nerve impulses, but the critical depolarization necessary for a dendritic action potential is well above the axon hillock threshold. The result is that slow potentials spread passively from the dendrites through the soma, and to the axon hillock region where action potentials are generated; each action potential produced at the axon hillock then sweeps (actively) not only along the axon but simultaneously back up the dendrites. Because there is a slight delay in this process, a recording made from the soma (which appears to behave like a dendrite) shows a peculiar looking action potential with an inflection on its rising phase (Figure 4-6). The first part (or A component) of this action potential, the part up to the inflection, reflects the passive spread of the axon hillock action potential into the soma and base of the dendritic tree; at the point of inflection, the soma and part of the dendrites give their own action potential (the B component), which, of course, appears larger to the electrode

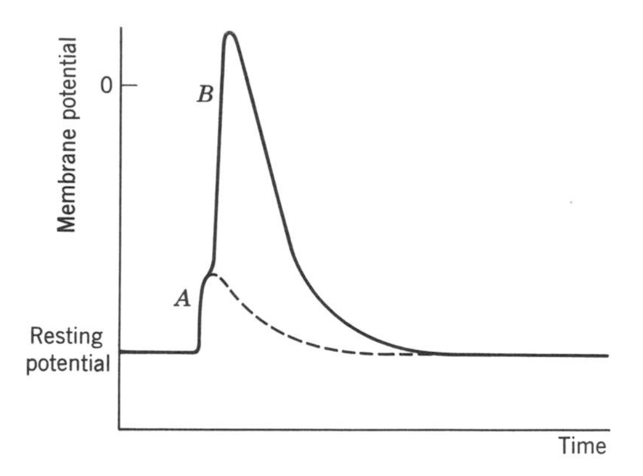

Figure 4-6. Action potential with inflection on rising phase. A and B components are indicated. The dotted curve represents what would have been seen if the soma and dendrites had been incapable of producing an action potential.

located in the soma than the one spreading passively from the axon hillock. In summary, then, the soma and part of the dendritic tree are probably electrically excitable in many neurons. However, by comparison to the axon hillock, the soma-dendritic membrane can be said to be relatively inexcitable, and the preceding description of neuron behavior based upon the assumption of (absolute) electrical inexcitability is not, in this context, inaccurate.

Thus far it has been assumed that if two PSPs occur simultaneously, their combined amplitude is simply the sum of their separate amplitudes. In practice this is not the case since the amplitude of a PSP itself depends upon membrane potential. For example, if a neuron is depolarized, its IPSPs are much larger than if it is hyperpolarized. In general, the more depolarized a neuron, the larger its IPSPs, and the more hyperpolarized, the smaller its IPSPs; if a neuron is sufficiently hyperpolarized, an IPSP will actually reverse direction and become depolarizing rather than hyperpolarizing (see Figure 4-7). In other words, an IPSP tends to drive the neuron's membrane potential toward a definite value, known as the **inhibitory equilibrium potential;** if the membrane potential is more depolarized than this inhibitory equilibrium potential, an IPSP will be hyperpolarizing, and if the membrane potential happens to be more negative than the equilibrium potential, an IPSP will be depolarizing. Finally, an IPSP will not alter a neuron's membrane potential if the membrane potential equals the inhibitory equilibrium potential. This last statement

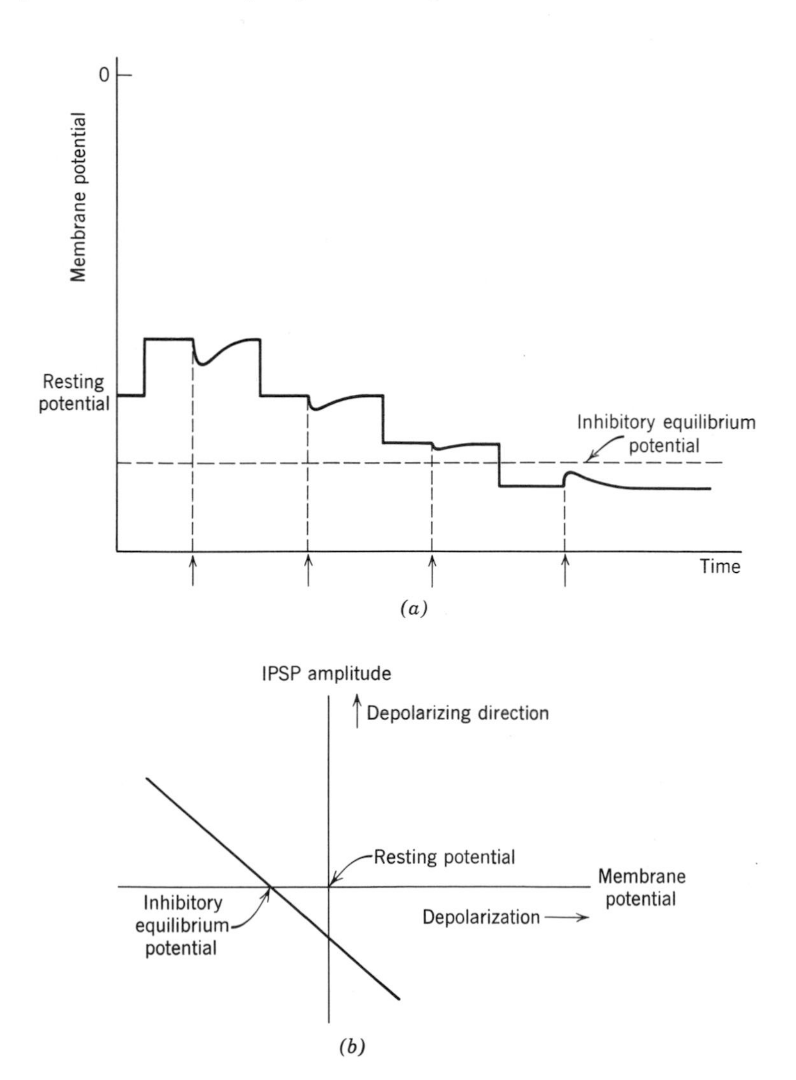

Figure 4-7. Dependence of IPSP amplitude on initial membrane potential. (*a*) Recording that might be obtained from a neuron in which IPSPs were produced at a number of different levels of membrane potential. Nerve impulse producing IPSP arrived at times indicated by arrows. The various values of membrane potential were produced by applying the appropriate voltage to an electrode within the neuron. (*b*) Curve relating IPSP magnitude to membrane potential.

may, in fact, be considered the definition of inhibitory equilibrium potential: that membrane potential for which activation of an inhibitory synapse causes neither de- nor hyperpolarization.

It should now be clear why IPSPs do not add linearly. Since IPSPs themselves change the membrane potential of a neuron, and since the magnitude of an IPSP in turn depends upon membrane potential (as shown in Figure 4-7), the size of two IPSPs occurring simultaneously will not usually be exactly equal to the sum of the sizes of each occurring separately. This phenomenon of nonlinear summation is well illustrated in the temporal summation of IPSPs (Figure 4-8). Suppose that IPSPs start to occur in a previously resting neuron at a constant rate, and further suppose that the peak amplitude of the IPSP is 1 millivolt at the resting potential. The first IPSP, at its peak, will hyperpolarize the neuron by 1 millivolt. Later IPSPs will add to the remaining hyperpolarization from previous IPSPs, resulting in temporal summation as described in Chapter 3. After a number of IPSPs have occurred, the neuron may have been hyperpolarized by a total of perhaps 5 millivolts. At this point the next IPSP to occur will do so in a neuron whose membrane potential is 5 millivolts more negative than the resting potential and as a result of the relation between IPSP magnitude and membrane potential, might hyperpolarize

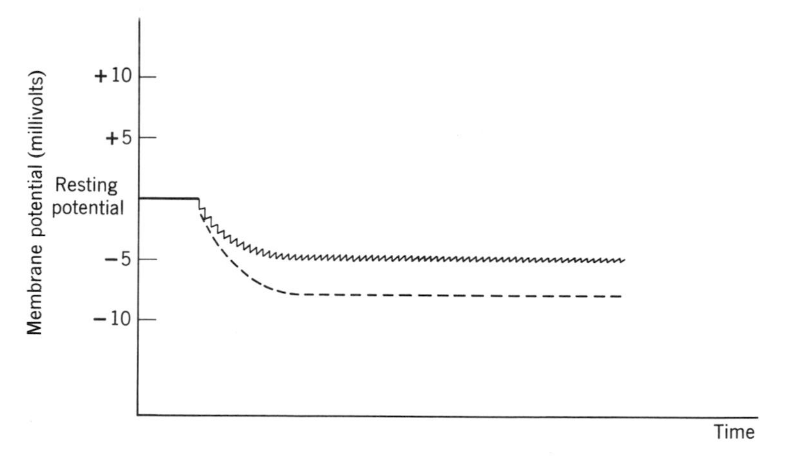

Figure 4-8. Nonlinear temporal summation of IPSP. The dotted curve gives the membrane potential expected if IPSP amplitude did not depend upon membrane potential.

the neuron by only perhaps an additional ½ millivolt (rather than the 1 millivolt for the cell at resting potential). Thus this later IPSP would add only half of the hyperpolarization expected on the basis of linear summation. In general, each succeeding IPSP would add a little less to the total hyperpolarization than the IPSP occurring just before.

This discussion immediately raises two questions. First, why were IPSPs described earlier as being hyperpolarizing PSPs, when they may either de- or hyperpolarize the neuron, depending on membrane potential? Second, why was the process of temporal summation described in terms of linear summation of IPSPs when they so obviously add nonlinearly? The reason, aside from the desire to postpone these complications until certain main points were made, is that the hypothesis of linear summation of IPSPs is often qualitatively, sometimes even quantitatively, accurate. This is so because in many types of neurons studied so far, with some important and conspicuous exceptions, the arriving excitatory and inhibitory PSPs balance in such a way that the membrane potential is often maintained within fairly narrow limits. Thus for these neurons, IPSPs are typically hyperpolarizing and seem to add approximately linearly as described earlier. In all but the very worst cases, the notion of (linear) temporal summation is generally useful in understanding qualitatively neuron behavior.

Like IPSPs, the size of an EPSP also depends upon membrane potential. For the EPSP, however, the direction of the dependence is opposite to that described for the IPSP. Thus EPSP amplitude decreases with depolarization and increases with hyperpolarization. Again there is a membrane potential for which an EPSP would cause neither a de- nor a hyperpolarization, the excitatory equilibrium potential. This equilibrium potential is in the neighborhood of zero volt. As with IPSPs, each EPSP in temporal summation has a slightly smaller effect than the one preceding, but again the amplitude of the EPSP is nearly constant over the range of membrane potentials normally seen in neurons. Thus within the normal limits of membrane potential, the linear addition of EPSPs in temporal summation described previously is often a sufficiently accurate approximation.

In summary, we can say that the axon has certain properties, passive spread of potential, threshold, all-or-none response to electrical stimulation and refractory period, which specialize it for information transmission. The dendrite serves, it has been said, as the neuron's receptive surface and has properties which suit it for this role: relative electrical inexcitability, passive spread of potential, and graded responsiveness to certain chemical transmitters. Action potentials have been contrasted with PSPs as the responses typical of axons and dendrites respectively. All these properties, in combination with neuron structure, enable the nervous system to carry out computations; the properties of axons permit the translation of a slow potential into impulse frequency and the rapid transmission of information in this form over relatively long distances; the properties of dendrites serve, through temporal summation of PSPs, to decode the nerve impulses into a replica of the slow potential from which they were generated; finally, all slow potentials thus transferred to a neuron's dendrites spread to the soma where they sum. This, in broad outline, is a view of nervous system function shared by many neurophysiologists. A number of questions, however, remain unanswered. Where do the slow potentials come from in the first place? That is, how does information enter (and leave) the nervous system? Even granting that the nervous system does compute by summing slow potentials, how are circuits of neurons arranged so that such computations result in the gross behavior of organisms? How is information stored within the nervous system? What are the biophysical mechanisms underlying the neuronal properties we have described? Although complete answers to these questions are not, of course, available—indeed, the accuracy of the slow potential theory itself is still under investigation—most of the remaining chapters will be devoted to presenting partial answers.

Chapter 5

INPUT AND OUTPUT: COMMUNICATION BETWEEN THE NERVOUS SYSTEM AND THE ENVIRONMENT

HOW INFORMATION ENTERS THE NERVOUS SYSTEM

In the preceding chapters we described a number of neuron properties and explained how these properties taken together provide the means by which the nervous system can carry out certain basic computations. A central notion of the previous discussion is that computation can occur by the summation of slow potentials; we have not, however, said where these slow potentials originate, nor have we mentioned their ultimate fate. In other words, the question of how information enters and leaves the nervous system has not been considered. This problem of communication between the nervous system and the environment is the subject of the present chapter.

To insure survival, an organism must have information about certain physical and chemical characteristics of his environment. A variety of specialized organs known as **receptors** are responsible for obtaining this information by detecting, for example, pressure changes, temperature, electromagnetic radiation of particular wavelengths, and concentrations of certain chemicals.

Generally, a particular receptor type detects only one physical or chemical characteristic of the environment and is relatively insensitive to others. For example, cells within the eye are exquisitely sensitive to light of the proper color and much less sensitive to pressure and temperature; pressure and temperature changes of the magnitude usually experienced by the eye are insufficient to stimulate these cells at all. In most cases, a receptor's sensitivity to distinct characteristics of the environment is sufficiently marked so that the receptor may be said in practice to be sensitive only to one type of stimulation. (This, in fact, is what we usually mean by "distinct.")

Although there are many types of receptor organs, each responsible for reporting to the nervous system the presence of a specific type of stimulus, it appears that their manner of reporting information is quite uniform. In each case, stimulus intensity is converted into a depolarization (quite analogous to the slow potentials described previously) which is in turn translated into nerve impulse frequency for transmission to neurons within the brain (Figure 5-1). In receptors, this depolarization reflecting stimulus intensity is known as a **generator potential**. Thus infor-

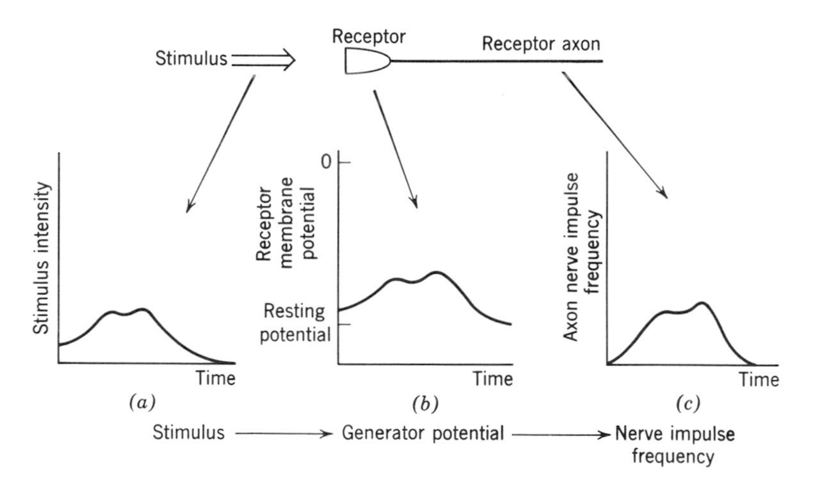

Figure 5-1. Entry into the nervous system of information about the environment. Stimulus intensity (light intensity, for example) (*a*) is translated into a generator potential by the receptor (a photoreceptor, for example) (*b*) which is in turn coded as nerve impulse frequency (*c*) for transmission to other neurons.

mation about the environment enters the nervous system through the conversion, or **transduction**, to use alternative terminology, of environmental properties into a slow potential. Although the end result of the transduction process is thought to be the same for all receptors (production of a generator potential), the exact nature of the process is, of course, quite different for each receptor type: a light receptor translates the brightness of a light shining upon it into a generator potential and has no special mechanism for producing generator potentials in response to pressure, whereas a pressure receptor has properties which permit pressure to be converted into a generator potential; the exact mechanism of the transduction process is not yet understood for any receptor, although some steps in the process are understood for certain types, most notably photoreceptors (light receptors).

Even though the mechanism by which stimulus intensity is converted into a depolarization is as yet only partly understood, several generalizations can be made about the process. It has already been stated that stimulus *intensity* is converted into a generator potential amplitude; if the experimenter uses the proper measure of stimulus intensity—usually "natural" measures such as concentration of a chemical, or energy of electromagnetic radiation—the quantitative relation between magnitude of depolarization and stimulus intensity is the same for most receptors. As an approximation, the generator potential is proportional to the logarithm of stimulus intensity (Figure 5-2).* It must be emphasized that this relation is only approximate, and, is particularly inaccurate for low stimulus intensities. Nevertheless, over wide ranges of stimulus intensity, the degree of conformity to a logarithmic stimulus-response relationship is often surprisingly close. More important than the exact mathematical description of the relation between stimulus intensity and receptor generator potential is the general form of the relation. For nearly all receptors, a fairly small change in stimulus intensity changes generator potential considerably at lower stimulus intensities, whereas a much

* The fact that generator magnitude is related logarithmically to stimulus intensity, together with the fact that generator potentials can be transferred from neuron to neuron through the nervous system with only a change of scale (linearly amplified or attenuated) presumably form the basis for the Weber-Fechner "Law" of Psychology.

$$dR = k\frac{dS}{S}$$

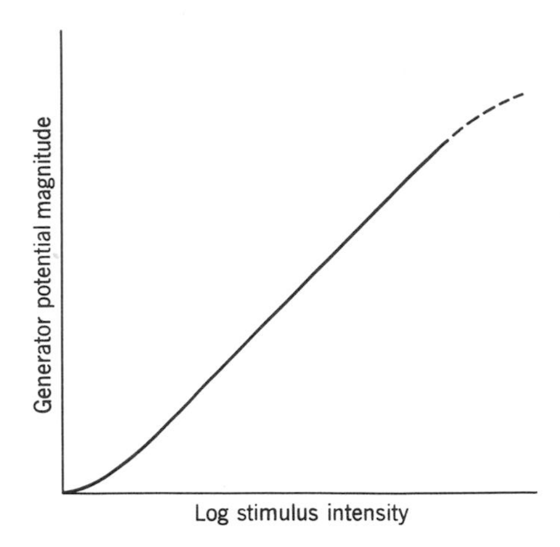

Figure 5-2. Relation between the logarithm of the stimulus intensity and generator potential magnitude for a typical receptor.

larger change in stimulus intensity is required to produce the same change in a generator potential at larger stimulus intensities. In other words, the slope of the relation between generator potential magnitude and stimulus intensity is progressively smaller for increasingly intense stimuli. Such a relationship between stimulus intensity and generator potential magnitude has the advantage of providing good sensitivity for *changes* in low intensity stimuli while at the same time permitting the receptor to operate over a large range of intensities (by decreasing sensitivity to changes in intense stimuli).

To further increase the ability to detect small changes in stimulus intensity, some receptors have mechanisms which enable them, in effect, to adjust the average stimulus intensity so that it falls within the sensitive range of the receptor. The human eye, for example, has at least two such mechanisms which serve this function. The first of these is the pupillary reflex which adjusts the pupil size according to the average light intensity and thus controls the average amount of light entering the eye. Although this mechanism is fairly rapid, it is not suitable for regulating the incoming light over a very large range since it is capable of no more than an approximately hundredfold modulation of the stimulus intensity.

The eye (as well as other receptors) has an additional mechanism for modulating the stimulus impinging upon it: this mechanism is most frequently referred to as the phenomenon of **adaptation.** If a photoreceptor is presented with a light of constant intensity, the generator potential caused by the light decreases in magnitude in the face of unchanging stimulus (Figure 5-3). In the eye, this decrease is thought to be the result of a diminished number of light-catching molecules involved in the first step of converting light intensity into generator potential. Since a smaller fraction of light is detected, the effect is as if the intensity of light falling upon the eye were decreased. As a result of adaptation, then, the intensity of light to which the eye is exposed is, in effect, decreased so that it now falls into a range where the eye is more sensitive to small changes in intensity. By using these mechanisms, the human eye can operate well over a millionfold range of stimulus intensities.

Just as there are both tonic and phasic neurons, so are there some receptors which produce a nearly constant generator poten-

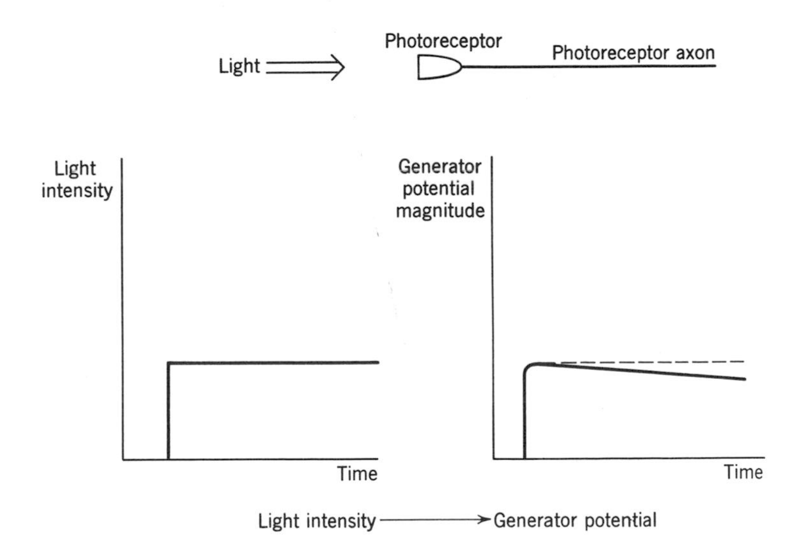

Figure 5-3. Adaptation in a photoreceptor. The light intensity indicated in the left graph results in the generator potential seen in the right graph. Response expected if there were no adaptation is indicated by the dotted line, while the actual response is shown in the solid line.

tial in response to an unchanging stimulus, and others whose generator potential declines rapidly in spite of a maintained stimulus. Receptors in the first category are referred to as *slowly adapting*, and those of the second kind are termed *rapidly adapting*. Rapidly adapting receptors respond best to stimuli of constantly changing magnitude, and it has been hypothesized that their function is to report information about changes in stimulus intensity to the nervous system.

This, then, is how information first gets into the nervous system: a stimulus is represented within a sensory neuron (receptor cell) as a generator potential that arises through the process termed transduction. Since the sensory cell is a type of neuron, it shares the properties of neurons, described in the first four chapters, and differs from them primarily in that its depolarization arises through the transduction process rather than through the temporal summation of ordinary PSPs. Like other neurons, many receptor cells code *generator potential* magnitude as impulse frequency and thus can transmit their generator potential to neurons within the nervous system proper. Since the sensory neuron behaves like other neurons, it also has a threshold; that is, there is a minimum stimulus intensity necessary for the production of nerve impulses. In other words, the receptor relays information about a stimulus only if the stimulus (and hence the generator potential) is above a certain, critical magnitude.

In summary, there are three things an organism can know about the various stimuli to which it is subjected: (1) the general nature of the stimulus (light, or temperature, or pressure, for example), (2) the intensity of the stimulus at each instant of time, and (3) the position of the stimulus relative to the organism. The nature of the stimulus can be recognized because there are many varieties of receptors specifically sensitive to distinct kinds of stimulation; by knowing which type of receptor is stimulated, the organism can classify the stimulus. Stimulus intensity can be determined by the organism, since the stimulus is, in effect, moved inside of a sensory neuron in the form of the generator potential, the magnitude of which accurately reflects stimulus intensity. Finally, the position of a stimulus can be determined because receptors are spread over the surface of the organism (and also inside, to give information about the internal environment) and respond only to a nearby stimulus. By knowing the location of the receptor, an organism can know the location of the stimulus.

HOW INFORMATION LEAVES THE NERVOUS SYSTEM

Although the nervous system can have an effect upon the organism's environment (either external or internal) in several ways, probably the most important mode of communication between nervous system and environment is through contraction of muscles. Without entering into details as yet, it may be said that muscles are organs which, in effect, transform nerve impulse frequencies into mechanical tension. The remainder of the chapter discusses how this is achieved.

A muscle is composed of a large number of very long cells (muscle cells) running in parallel for the length of the muscle (Figure 5-4). The muscle cells contain longitudinal fibrils, and these in turn have a fibrillar fine structure which we shall not consider here. Each muscle cell receives a number of axons (originating from neurons within the central nervous system known as motor neurons) which terminate in a structure known as the neuromuscular junction; this structure is similar to synapses seen within the nervous system itself and is, in fact, a type of synapse.

Muscle cells share many electrical properties with neurons: they exhibit resting potentials, have the property of passive spread of potential along their length, and if depolarized give action potentials much like those seen in neurons. In addition to these electrical properties, muscle cells have a mechanical property not seen in neurons: a shortening of the muscle cell accompanies a muscle action potential, and if the shortening is opposed by an external force, the muscle develops considerable tension. Several observations cause us to believe that the muscle action potential is normally the cause of the contraction. The mechanism by which an action potential results in a contraction, the excitation-contraction coupling mechanism, as it is known, is not yet well understood. The process is, however, well understood on the descriptive level, and this is sufficient for our present purposes.

If a muscle cell is constrained so that shortening cannot occur

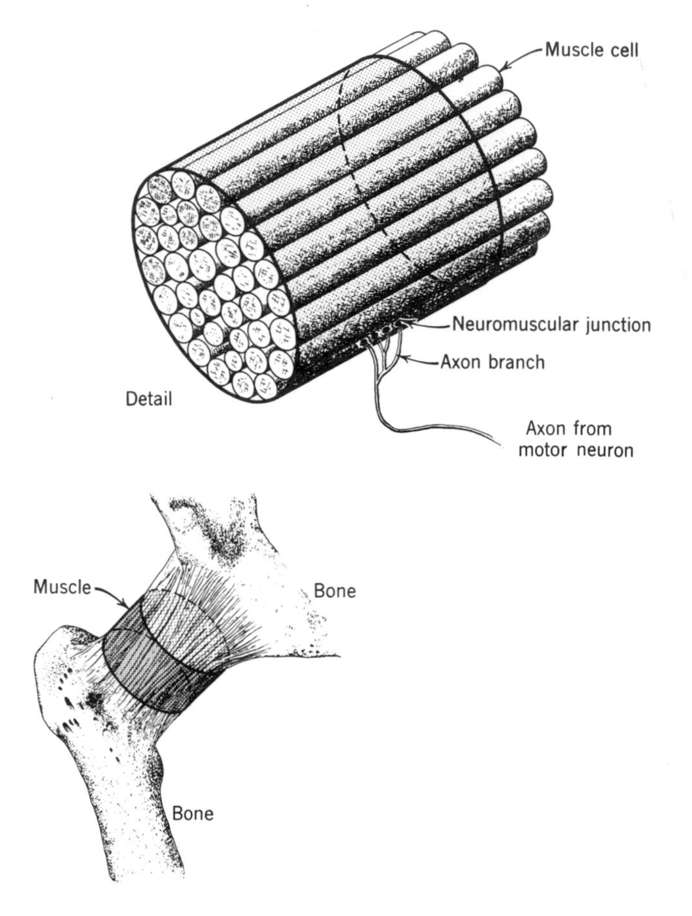

Figure 5-4. Schematic cross-section through a muscle showing the individual muscle cells (muscle fibers) and the innervation of one muscle cell.

following an action potential, and if the tension developed by the muscle cell is measured as a function of time, tension rises rapidly (as a concomitant of the action potential) and then returns slowly to the resting level; the shape of the process is reminiscent of that of the postsynaptic potential (Figure 5-5*a*). The event just described is known as a **muscle twitch** or a **twitch contraction**. If a second muscle twitch is initiated before the tension from a preceding twitch has returned to the resting level, summation of tensions is observed (Figure 5-5*b*). This summation of tension from sequential muscle twitches is analogous to the temporal summation of PSPs seen within neurons and, as will be seen, serves a similar function.

No mention has been made yet of how action potentials (and consequently twitch contractions) are ordinarily initiated in muscles. As might be expected from the structural similarity between the neuromuscular junction and synapse, an action potential arriving at the neuromuscular junction causes, in the muscle cell, a potential much like an EPSP which in turn triggers a muscle action potential. Thus it should be clear that a slow potential in a motor neuron (a neuron sending its axon to a muscle) can be transformed into nerve impulse frequency and the impulse frequency can be decoded into tension at the muscle through the temporal

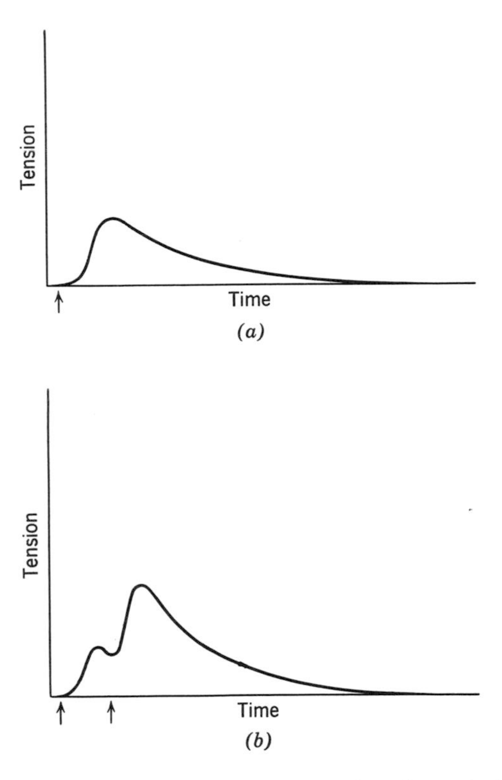

Figure 5-5. Muscle tension resulting from nerve impulse arrivals at the neuromuscular junction. (*a*) Twitch contraction resulting from a single nerve impulse. (*b*) Temporal summation of tensions from two successive twitches. Arrows indicate the times at which nerve impulses arrive.

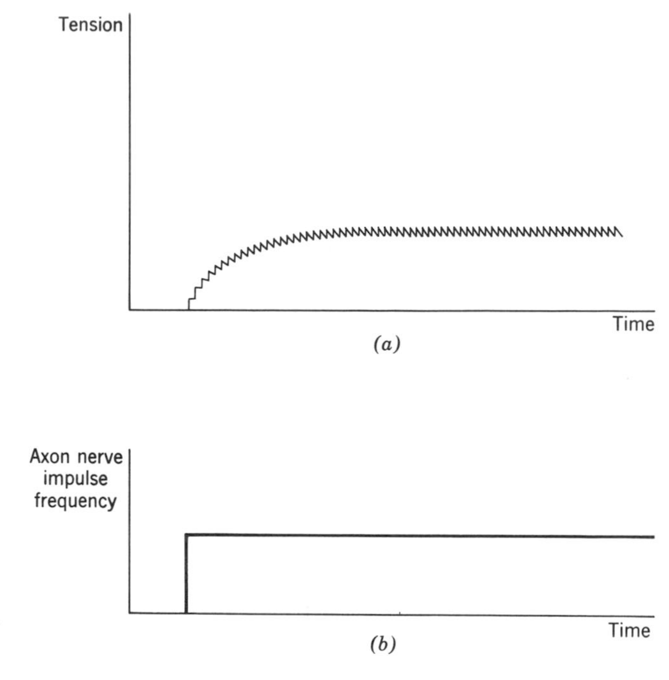

Figure 5-6. Summation of muscle twitches to produce a maintained tension. (*a*) Temporal summation of muscle twitches. (*b*) Nerve impulse frequency causing the muscle twitches.

summation of muscle twitches that are usually relatively prolonged, lasting perhaps hundreds of milliseconds. Because the muscle's PSPs are relatively shortlived (perhaps 10 milliseconds) and individually large enough to trigger muscle action potentials, decoding of the frequency message occurs through the summation of twitches and not by the summation of muscle PSPs.

Since muscle cells exhibit an analog of temporal summation, one might inquire if they also show a process analogous to spatial summation. The whole muscle is so structured that something very similar to spatial summation of tension does in fact occur. It should be recalled that a muscle is made up of many parallel muscle cells extending the length of the muscle. These individual muscle cells have a common attachment at either end of the muscle (to bones of a limb, for example) and thus the tension developed by the individual cells is summed to produce the total tension developed by the muscle. Individual muscle cells generally have separate innervation and therefore operate as independent units; this means that the contractions made up by summation of twitches of each muscle cell in an entire muscle are added together to form a grand contraction much as the slow potentials arising at separate synapses on a neuron can be added together to form a grand slow potential. Thus the total tension developed by a muscle is the sum of the tensions developed by the individual muscle cells, and these, in turn, depend upon the frequency of nerve impulses arriving at the muscle cell.

Chapter 6

COMPUTATION IN A SPECIFIC NEURAL CIRCUIT

Neurophysiology is in the somewhat paradoxical position of knowing much more about the properties of neurons than about the principles of organization of neural circuits employing those properties. In fact, for no neural circuit is there a complete and detailed understanding of the circuit's overall behavior in terms of properties of the constituent neurons and their interconnections. Nevertheless, in a few specific situations, understanding of a neural circuit, although not complete, is quite extensive. One such circuit is the neural network in the eye of the Horseshoe crab, *Limulus*. In this chapter we shall use the *Limulus* eye neural network to illustrate the sort of computations that can be carried out by relatively simple neural circuitry. The *Limulus* eye has been chosen for discussion not only because it is an extensively studied system, but also because it exhibits some principles of neural organization apparently quite common in the mammalian nervous system. Since our main concern is with illustrating the role individual neurons play in neural circuits, we shall present a rather schematic version of *Limulus* eye physiology.

The *Limulus* eye is like the human eye in that an image of the environment is formed on a two-dimensional array of photoreceptors, but it is quite unlike the human eye in the optical method of image formation. Mammalian eyes, cameralike, have (in effect) a single lens which focuses an image of the outside world on the photoreceptor mosaic. *Limulus,* on the other hand, like the honey bee, has a compound eye in which each photoreceptor is provided with its own lens for concentrating light from a particular restricted part of the environment onto that receptor. Each receptor then reports the (average) light intensity in a small patch of the external world to the animal's central visual processing station; that patch of environment reported on by a photoreceptor is known as the photoreceptor's visual field. The patterns of light and dark around *Limulus* are thus translated into generator potentials, coded as nerve impulse frequency, and finally transferred to an area known as the optic lobe where further analysis, such as pattern recognition, is carried out. Ideal operation of such an eye is illustrated in Figure 6-1.

The behavior of *Limulus* eye departs from the idealization illustrated in Figure 6-1 in several significant ways. One of the important departures is that the visual fields from adjacent receptors do not exactly abut as illustrated, but rather overlap to quite a considerable extent as shown in Figure 6-2*a*. The effect of this overlapping of visual fields is perhaps best illustrated by considering how an edge, that is, a sharp step in brightness, is represented in the generator potential magnitudes (Figure 6-2*b* and *c*). Ideally, of course, the edge would be represented in the receptor mosaic as a sharp difference in generator potential magnitude between the neighboring receptors reporting on the light intensity in the part of the environment where the edge occurred. In practice, however, because of the receptors' overlapping visual fields, there is not sharp transition in generator potential magnitude between neighboring receptors, but rather a blurring of the edge as represented by the generator potentials. This blurring occurs because receptors which ideally should be looking only at the environment near the edge actually see part of the environment on the other side of the edge. That is, because of the overlapping visual fields, a receptor which should report seeing a relatively dim light in practice gathers some of its light from the brighter part of the environment, and thus (since a receptor senses the average light intensity over its entire visual field) reports

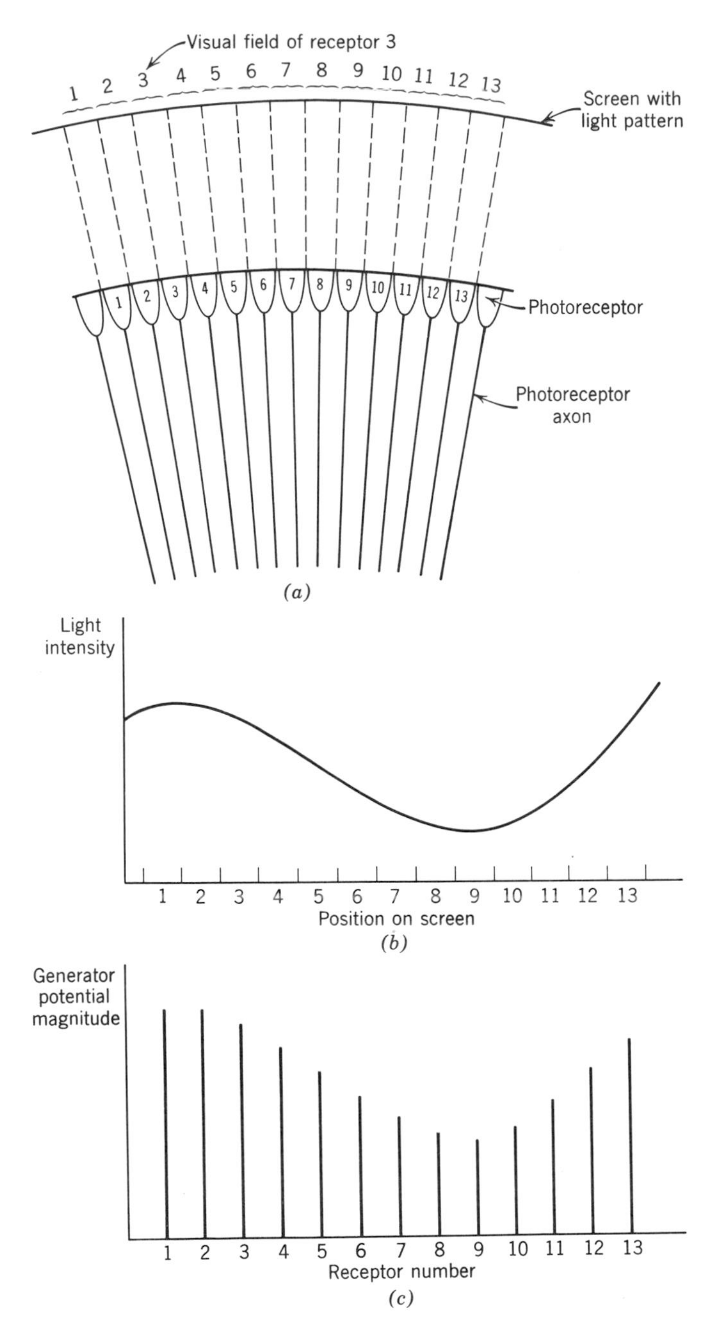

Figure 6-1. Ideal operation of a compound eye. (*a*) Schematic cross section of a compound eye looking at a light pattern on a screen. Brackets next to screen indicate the visual fields of the receptors. (*b*) Light intensity as a function of position on the screen. The numbers on the abscissa indicate position of the visual fields for the various receptors illustrated in (*a*). (*c*) Generator potential magnitudes produced in the receptors by the light intensity pattern illustrated in (*b*).

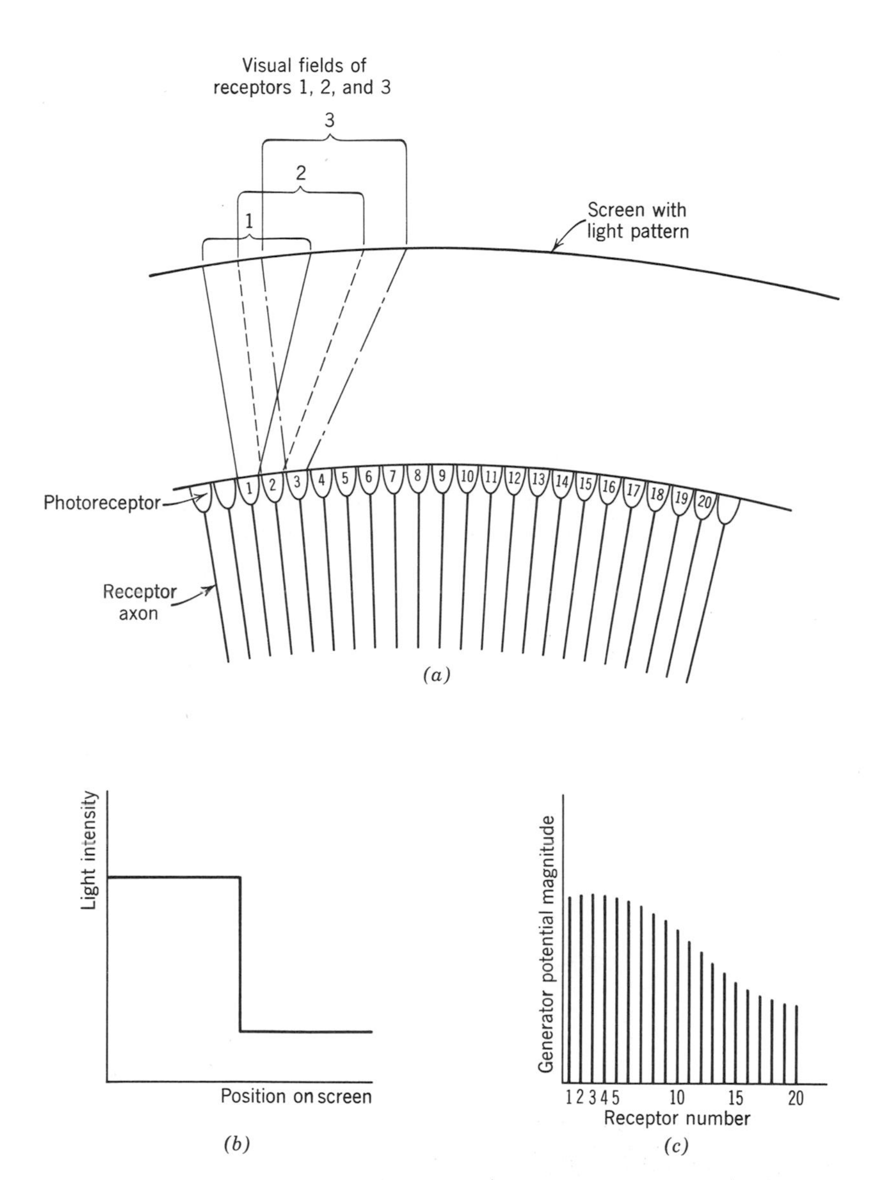

Figure 6-2. Actual operation of a compound eye. (*a*) Overlap of visual fields. (*b*) Light intensity pattern on screen viewed by eye. (*c*) Generator potentials produced in the receptors by the light intensity pattern shown in (*b*).

seeing a light of intensity intermediate between the bright light on one side and the less intense light on the other. Altogether, then, this overlapping of visual fields decreases the resolving power of the eye.

Through the use of neural circuitry, however, *Limulus* manages not only to compensate for the overlapping of visual fields, but actually to enhance somewhat the contrast at edges. This is accomplished by a simple and elegant process known as **lateral inhibition:** each receptor in *Limulus* inhibits its neighbors.

When a *Limulus* eye receptor codes its generator potential as nerve impulse frequency for transmission to the optic lobe, it also sends this frequency-coded generator potential to neighboring receptors, where the message is decoded by the temporal summation of IPSPs. That is, a receptor in effect subtracts a portion of its generator potential from the generator potentials in the neighboring receptors. Only a fraction is subtracted because the generator potential is attenuated when it is decoded in the neighboring receptor; this attenuation occurs through the operation of the factors enumerated on page 57 of Chapter 4. Exactly how much a generator potential from one receptor is attenuated when it is decoded in a neighboring receptor depends on the distance between the two receptors: the greater the distance, the more a generator potential is attenuated. The type of relation between attenuation factor and the distance between the receptors is indicated in Figure 6-3. Such a relation (Figure 6-3) is thought

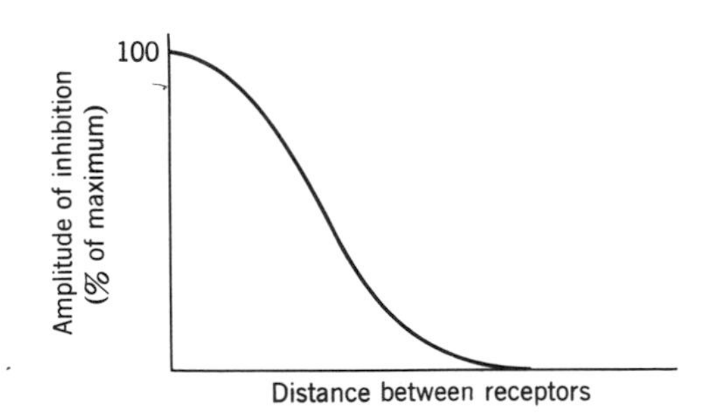

Figure 6-3. Amount of inhibition caused in one receptor by another as a function of distance between the two receptors. Strength is indicated as per cent of inhibition in the nearest neighbors (those receptors most strongly inhibited).

to be a consequence of the anatomical connections between the *Limulus* receptors. As indicated in Chapter 4, the amplitude of the slow potential decoded from nerve impulse frequency may depend on a number of different factors; in the *Limulus* case, the attenuation as a function of distance may be presumed to result primarily from the number of inhibitory synapses one receptor contributes to another. The axon from a particular receptor is thought to branch many times before it synapses, its near neighbors receiving a relatively large number of these branches and more distant neighbors considerably fewer.

Before considering the use of lateral inhibition in *Limulus* eye, we must clarify a statement appearing earlier in this chapter. Thus far we have spoken as if a receptor's generator potential were coded as impulse frequency for transmission to other neurons. Because of the existence of lateral inhibition, it is not the generator potential alone, but rather the total slow potential, consisting of generator potential minus any inhibitory potentials, which determines the frequency of the receptor. Despite the fact that this distinction is important, we shall on occasion continue to speak loosely of a receptor subtracting a portion of its generator potential, rather than its total slow potential from generator potentials in neighboring receptors. For certain stages of qualitative arguments, it is unnecessary to maintain the distinction between generator potential and total slow potential because the generator potential is usually the major component of the receptor's slow potential.

Having described the phenomenon of lateral inhibition, we shall consider its use to *Limulus*. The effect of lateral inhibition is quite easy to state: it serves to compensate for the overlapping visual fields of the receptors and, furthermore, to emphasize or sharpen the edges between areas of different brightness. Understanding in qualitative terms exactly how this final effect occurs is somewhat more difficult. We begin by showing the eye an edge, that is, a step change in intensity, and consider as before the generator potentials resulting from this pattern of light intensities (Figure 6-4). Because of the overlapping visual fields, the pattern of generator potentials shows a loss of sharpness of the edge: there is a blurring of the sharp change in light intensity in the generator potentials' representation of the pattern. To a first approximation, these generator potentials are coded as nerve impulse frequency and transmitted to neighboring receptors for decoding into inhibitory potentials. The final slow potential in each receptor

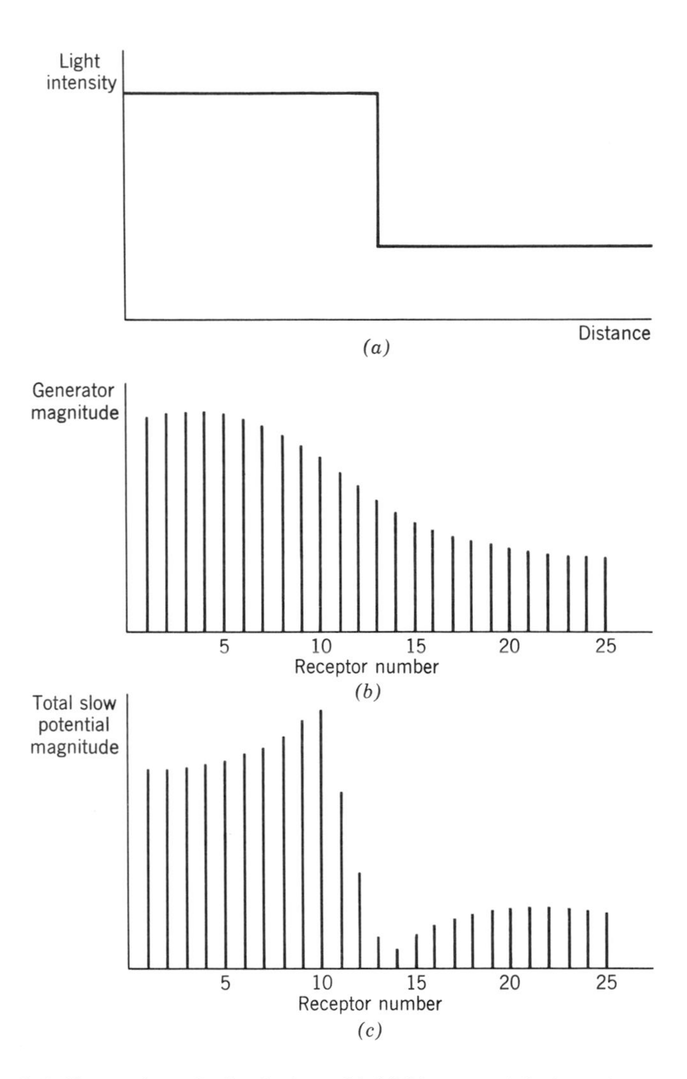

Figure 6-4. Sharpening of edge by lateral inhibition. (*a*) Light intensity pattern on screen placed before the eye. (*b*) Generator potentials produced in the receptors by the light intensity pattern in (*a*). (*c*) Total slow potentials produced by "correcting" the generator potentials with lateral inhibition.

will be the algebraic sum of the generator potential in that receptor and all of the inhibitory potentials transferred there from neighbors.

Starting with receptor 1 on the left of Figure 6-4, we see that this receptor will have a relatively large amount subtracted from its generator potential, since all its neighbors have large generator potentials (all cells in this part of the eye are looking at a bright region). The total effect of inhibition from all neighbors produces in receptor 1 a grand slow potential of, for instance, the magnitude indicated in Figure 6-4*c*.* A similar argument holds for receptor 2; it sees a bright light (and so has a large generator potential), and it is surrounded by other receptors looking at bright regions. We turn now to receptor 25 on the other side of the eye. This receptor has a small generator potential to start with, but not too much is subtracted from it since the neighboring receptors also have small generator potentials (they too are looking at dim regions). A similar argument holds for receptor 24 since it sees a dim region and is surrounded by neighbors also looking at dim regions. It should be emphasized at this point that we have been making use repeatedly of the fact that the largest inhibitory effects experienced by a receptor are those arising in the near neighbors. We need not, in general, consider the effects of the more distant neighbors, but may estimate the total inhibition by examining the generator potentials in, for example, the next and the next-but-one receptors.

The situation is quite different from the one we have considered so far for receptors near the transition from light to dark. Taking receptor 10 next, we see that it has a large generator potential, and that it experiences less total inhibition than receptor 1 or 2. Receptor 10 receives strong inhibition from its left (these cells are looking at bright regions), but receives much less inhibition from the right since the cells on its right have smaller generator potentials. The total slow potential, then, is smaller than the original generator potential but larger than the grand slow potential in receptor 1 or 2 (which receive strong inhibition from both sides). Next receptor 14 has a small generator poten-

* To avoid end effects, we have assumed that the light pattern and the eye stretch to infinity on both sides.

tial, receives a relatively small amount of inhibition from its right (the cells there are in the dim region), but receives a large amount of inhibition from its left since receptors on that side have large generator potentials. The net result is a slow potential smaller than that in cell 24 or 25 even though these receptors have the slightly smaller generator potentials. Carrying out similar arguments for the other receptors, it can be seen that the final result would be a pattern of slow potentials, that is, generator potentials corrected by inhibition from neighbors, as indicated in Figure 6-4*c*. Not only has sharpness been restored to the receptor mosaic representation of the edge, but the eye's representation of the environmental light pattern makes the transition region even more prominent. When looking at step change in intensity, then, what the eye sees is an edge with a brighter band at the transition region on the bright side and a dark band at the transition region on the dim side. These apparent light and dark bands at the transition from a bright to a dim region are known as **Mach bands,** and form the basis for a number of dramatic optical illusions experienced by humans. The usual explanation given for the phenomenon in human vision is roughly the one given here for the *Limulus* eye.

A further word should be said about the preceding argument. We supposed that each of the generator potentials in an array of receptors was transmitted to neighboring cells where a fraction of it was subtracted from whatever generator potential was there. Considerations of this sort lead us to a set of slow potentials which may be thought of as "corrected generator potentials." Since, however, it is these "corrected generator potentials" which actually determine the output of the receptor, we should carry out the argument a second time to arrive at a set of "doubly corrected generator potentials." But these, too, should again be corrected, and so on indefinitely. Although it may not be intuitively clear, a quantitative analysis reveals that after an infinite number of corrections and recorrections the generator potential finally ends up such as we predicted from the simple analysis carried out in the previous paragraphs.

This brief chapter has illustrated that a few thousand neurons in even a very simple circuit can perform computations useful to the organism. In *Limulus,* for example, simply having a receptor inhibit its neighbors compensates for imprecision in the eye's optical apparatus and further serves to enhance contrast at borders. It should not be difficult by extrapolation to imagine the computational power available from a brain containing ten billion neurons interconnected in complex circuits.

Chapter 7

LOCALIZATION OF FUNCTION: THE STUDY OF CELLS IN AGGREGATE

A complete understanding of nervous system operation would require a detailed knowledge of how each neuron affects others. Since even comparatively simple animal brains contain millions of nerve cells, a neuron by neuron analysis of an entire nervous system is obviously impossible simply because of sheer numbers of cells, all other technical difficulties aside. Nevertheless, it is possible to make progress in understanding total brain function without the necessity of studying every cell, or even without examining the behavior of any individual neurons. Because of the anatomical arrangement of neurons in a brain, numerous nerve cells may be investigated simultaneously, and in this way much can be learned about how the brain operates.

In the vertebrate brain a given function is generally not performed by a single neuron, but rather by a relatively large number of neurons all doing about the same thing at the same time. Such redundancy of operation greatly increases the reliability of nervous system function, and has certain other advantages as well. Because of the orderliness of brain structure, it might be expected that neurons with the same function would be located in the same brain region. This, indeed, is generally the case. The notion that nerve cells with the same tasks are close together, and further that cells in a particular region of the brain are involved in the same functions, is known as the doctrine of **localization of function.** Although this doctrine has been much abused in the past, it is now taken simply as the physiological correlate of the orderliness of brain structure. Assuming that localization of function is not taken in too literal a form, it provides the rationale for the study of cells in aggregate: to the extent that it is accurate, the doctrine of localization of function implies that one may study a particular function of the nervous system by examining cells in specific regions of the brain. This, in turn, offers the possibility of studying groups of cells simultaneously rather than examining them individually. For the remainder of this chapter, we shall be concerned with the techniques developed for unraveling nervous system function by studying aggregates of cells.

LESION TECHNIQUES

One of the oldest and still most important techniques for studying neurons in aggregate is to destroy all of the cells in one small region of the brain and then to observe how brain operation is altered as a result of the **lesion.*** From the nature of the change in brain function, together with other types of evidence, it is often possible to assign a role in nervous system operation to neurons in the area destroyed. Quite frequently, some function of the brain will be lost completely: for example, the animal may be blind, or deaf, or unable to move a limb after the brain area has been destroyed. In such cases of a deficit following the loss of cells in a specific area, it is usually inferred that those cells were intimately involved in performing the lost function. Some lesions do not cause deficits, but rather result in exaggerated performance of some operation. For example, destruction of cells in a particular area of the hypothalamus leads to almost continuous eating and consequently to enormous increases in body weight. Such an increase in a type of behavior is known as a **release phenomenon,** and is taken to indicate that the cells which were

* *Lesion* is used as a noun, to indicate the area of destroyed cells, and also as a verb to denote the act of destroying cells.

destroyed normally serve to inhibit other neurons responsible for the behavior which has become exaggerated.

A number of different techniques are used to destroy neurons. If a fairly large lesion is to be produced, particularly in animals with larger brains, the area of brain under study is simply removed from the animal surgically. This technique has numerous advantages, primarily in the ease with which large lesions are made and in the accuracy of placing certain lesions, but suffers from the disadvantage that lesions often cannot be placed in the center of the brain without undue damage to surrounding structures. It is also difficult to selectively destroy very small areas of the brain by simple surgical removal.

A second procedure frequently used to produce lesions is heating the area to be destroyed until cell proteins are denatured and the tissue is coagulated; the region is literally cooked, just as one coagulates an egg white with heat. Since very small regions can be destroyed in this manner by passing a relatively large current through an electrode, the technique is particularly well adapted for producing minute lesions in the depth of the brain. All brains of a particular animal species conform quite closely to the same map, and the structures within the brain bear a fixed relationship to certain external landmarks on the animal's skull. It is possible, therefore, to insert an electrode into the depths of the brain and destroy a preselected nucleus with considerable, although not complete, accuracy. Disadvantages of this technique, aside from the obvious and often serious difficulty of controlling the exact position and size of the lesion, are first that the shape of the lesion is usually roughly spherical (which is desirable only when one wishes to destroy a spherical region) because the heating is about the same in all directions, and second that the area in which cells are destroyed does not have distinct borders. Nevertheless, this lesioning technique is probably the one most frequently used in experimental work.

Occasionally lesions are produced by injecting toxic chemicals into localized brain areas; alcohol, for example, has been used for this purpose. However, there is usually no advantage of chemical, over heat-produced lesions, and because the method is somewhat more difficult to use with precision, it is not frequently employed. On the other hand, temporary lesions produced by chemicals (such as local anesthetics) do have an advantage over the techniques described previously in that the effect is a reversible one. Reversible lesions produced by injection of pharmacological agents, and also by local cooling of the brain, are being used more frequently now; these techniques will probably be of considerable importance in the future, since some types of experiments require that the lesion not be permanent. Finally, three other lesioning techniques may be mentioned briefly. In order to produce lesions in the depths of the brain without the necessity of introducing an electrode through brain tissue which the experimenter does not wish to damage, focused ultrasound and ionizing radiation have both been used. These techniques are frequently unreliable, and not very useful for the production of very small, well circumscribed lesions (which are often desirable in experimental work); in certain special situations, however, the methods have proved valuable. A third occasionally used lesioning technique makes an area of the brain inoperative by forcing it to behave in some way incompatible with normal functioning; this is most often accomplished by applying such a large stimulus to cells within a particular region that any normal input they might receive is inconsequential. Although reversible "lesions" are produced, such a technique is so fraught with pitfalls that interpretation of results, something difficult enough in the very best lesion experiments, is next to impossible.

STIMULATION TECHNIQUES

If preventing neurons from producing action potentials has one effect, causing them to produce action potentials should have the opposite effect, and often it does. The converse of lesion experiments, then, is to cause neurons in an area to become excessively active by stimulating them. In stimulation experiments one infers the function of neurons from the effects observed during their increased activity. The logic involved is dangerously close to the notion that if one pill is good for a disease, five pills should be five times as good; nevertheless, gross stimulation, carefully used, is one of our more valuable techniques.

Although a region of brain tissue can be stimulated to greater activity in a number of different ways, only two methods are generally useful: electrical stimulation and chemical stimulation. Electrical stimulation involves passing an appropriate electric current through a region of tissue and thereby causing neurons in the area to be depolarized. Although it may not be immediately obvious that such a current could depolarize neurons locally, indeed understanding the process quantitatively is difficult, it is nevertheless known from experiments that it can. Since the effect of appropriate stimulation is a depolarization, one can apply the knowledge gained from intracellular depolarization experiments to better interpret results of experiments involving electrical stimulation.

Chemical stimulation generally depends upon the fact that transmitter substances applied locally to the brain have the same effect upon dendritic membranes as the same transmitter applied naturally by release from axon terminals.* In theory, this technique might be extremely useful and selective, since injection of a particular transmitter into a region might be supposed to affect only dendritic area normally receiving that transmitter. There are indeed cases where such specificity is known, but more usually a large variety of chemicals will alter the behavior of any neuron to which they are applied. The situation is still more complicated by the fact that only a very few transmitter substances have been isolated and identified chemically; thus when one obtains a specific effect from using some chemical, it is very difficult to interpret the results in terms of normal brain function. It appears now that the technique of local application of chemicals will fulfill its promise as a useful method for gross stimulation only when we more fully understand transmitter pharmacology at the level of the single neuron.

RECORDING TECHNIQUES

The technique of gross electrical recording from a number of neurons depends upon the fact that action potentials and PSPs

* A very useful alternative is to apply substances which render more effective the transmitter released naturally. For example, it is sometimes possible to inhibit the enzyme which is responsible for destroying a transmitter, thus, in effect, selectively increasing any input to the neurons which involve that transmitter.

cause small voltage changes in the immediate vicinity of the active neurons (this matter is treated more fully at the end of Chapter 10). Thus a 100 millivolt action potential might result in a ½ millivolt fluctuation just outside (at a distance of perhaps 10 micra) the neuron of cell body. This extracellular voltage fluctuation would be smaller at greater distances, falling to undetectable levels at a distance of perhaps 50 micra. There are, of course, similar extracellular voltage changes accompanying PSPs which may be even smaller than the extracellular signs of the action potential.

For reasons which are as yet unclear, most areas of the brain exhibit spontaneous rhythmic fluctuations in voltage which can occasionally be correlated with altered states of brain function. For example, there are striking differences in the **electroencephelogram** (usually abbreviated to **EEG**), the term used for a recording of these voltage fluctuations, between a sleeping and wakeful animal. Aside from certain important applications in clinical medicine, however, the EEG has been of only limited usefulness. On occasion, a particular experimental manipulation will lead to a dramatic change in the form of the EEG in a very limited region of the brain, in which case it may be possible to infer something about localization of a particular function. Although specific EEG changes of this sort can be very helpful, the usefulness of the EEG as an analytic tool is severely limited by the lack of understanding of its cause and meaning for normal brain function.

The extracellular voltage fluctuations associated with PSPs and action potentials are very small indeed when the recording electrode is even a short distance from the active neurons. Nevertheless, if a large number of neurons contained within a small region of the brain are simultaneously active, that is, if all of their membrane potential fluctuations are nearly synchronous, appreciable voltage changes may be recorded throughout the region containing these neurons. This fact provides the rationale for a valuable method known as the **evoked potential technique:** when the inputs to a group of neurons are activated simultaneously, the response of the neurons (assuming the input has the same effect on each) is synchronous and there is an associated voltage fluctuation in the vicinity known as an **evoked potential.** Even if an evoked potential is complicated and its exact mode of production unknown, its very existence may be of considerable significance. For example, if stimulation in one region of the brain results in an evoked potential in another region, we may conclude that neurons in the first region eventually synapse (perhaps via a number of other neurons) with cells in the second. Since the size of extracellular potentials decreases rapidly with distance from the active neuron, an evoked potential is largest in the immediate vicinity of the active neurons, serving to pinpoint the brain local containing the activated neurons. Thus new pathways (that is, functional connections from one brain region to another) may be discovered by the systematical moving of stimulating and recording electrodes until locations are found which yield evoked responses. In fortunate circumstances, enough is known about the brain area under study that it is possible to interpret the form of the evoked potential in terms of membrane potential changes in a particular group of cells, and thereby to discover new properties of these cells and their connections. In the ideal situation, then, the evoked potential technique is one of our more important analytic tools; often, it is a great technical convenience, since it permits a group of cells to be located by a recording electrode in the depths of the brain (by stimulating a tract known to evoke a potential in the neuron group and moving the recording electrode until that potential is recorded). Unfortunately, however, we frequently do not know enough other facts about the brain region under study so that the form of the evoked potential can be understood; this together with certain technical hazards, make the evoked potential technique quite unreliable in many situations.

There is one final recording technique which in a sense is intermediate between the gross recording just described and the use of an intracellular microelectrode (as described in the first five chapters). It is possible to record the extracellular signs of action potentials from a single neuron, or from small numbers of neurons when the electrode is placed in proximity to the active cells. The voltage fluctuations thus detected are often small, but they are nonetheless sufficiently large to indicate at least the occurrence of action potentials. **Extracellular unit recording,** as this technique is called, does not give as much information as intracellular recording, but it is technically much easier and does have certain advantages. Sometimes it is useful to be able to examine the behavior of a few neurons simultaneously; furthermore, the extracellular electrode typically causes less damage to a neuron than penetrating the cell membrane with an electrode, so that extracellular recording may interfere less with the behavior the neurophysiologist is studying. Because it is relatively easy to use, extracellular unit recording is at present the most widely chosen method of investigating the behavior of single neurons.

ILLUSTRATION OF THE USE OF STIMULATION, LESION, AND RECORDING TECHNIQUES

A physiologist who studied hearing in fleas decided to give a public demonstration of his results at a meeting of fellow scientists. He used a flea that was trained to jump whenever the verbal command "jump" was given, and after convincing his observers that the flea would indeed jump when told, he removed all the flea's legs (very carefully, so as to cause no other injury). Once the legs had been removed, the flea would not jump, no matter how loudly the command "jump" was given. "You see," said the physiologist, "the flea can no longer hear. I have proved that fleas hear with their legs."

The use of these various techniques can best be illustrated by specific examples of the results from an investigation of a particular aspect of nervous system function. We have chosen the hypothalamic appetite **centers** * for this illustration, despite the fact that interpretation of many of the data is still in dispute.

If a lesion is made in the region of the ventromedial nucleus of the hypothalamus (Figure 7-1, region marked "presumed

* A center, loosely speaking, is a brain region where a particular function is localized. The term has been and continues to be much abused because we have a tendency to go too far beyond our experimental observations, and because our current notions of brain function are still in a primitive stage of development. In the following discussion, *center* should be taken to indicate no more than the existence of a nervous system location, the proper manipulation of which is correlated with changes in certain observables. Although this is the strict definition of center, we shall try to name our centers in such a way that violence is not done to the words used in the naming; that is, we try to pick names for centers which will, when we know more about brain function, turn out to be happy choices. Thus if a lesion in a particular structure led to cessation of respiration followed by death, we should prefer to term the structure a *respiration center* rather than a *death center*.

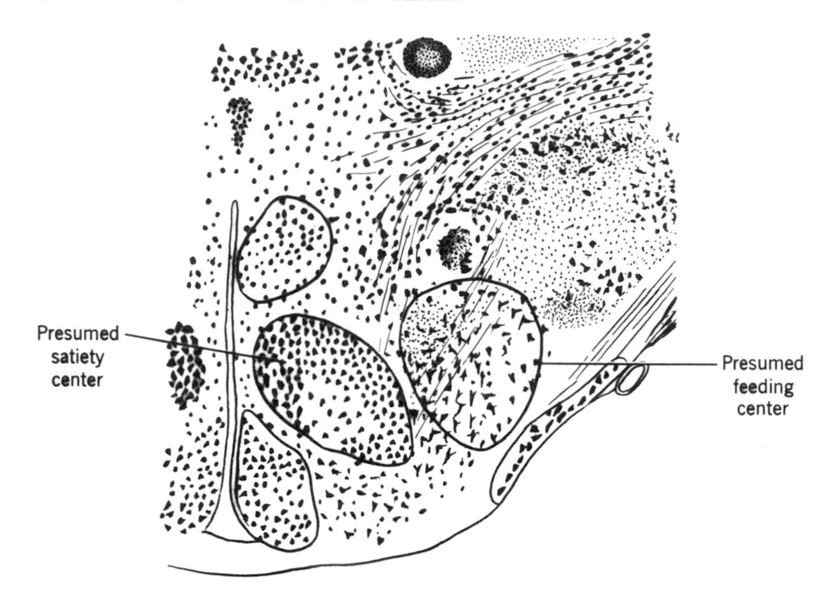

Figure 7-1. Cross section through the rat hypothalamus at the level of the ventromedial nucleus. (After Figure 5, W. J. S. Krieg, *J. Comp. Neurol.*, **55**:33, with the feeding and satiety centers indicated.)

satiety center"), the animal subsequently exhibits an enormous appetite. Such animals will eat almost continuously and have been known to quadruple their body weights. Occasionally, they will literally eat themselves to death in a very short time. Since removal of the medial hypothalamus results in increased eating, we infer that the normal function of this area is to limit food intake; the increased appetite is a release phenomenon, and the ventromedial area is inhibitory to some other area actually responsible for initiating eating. This region, apparently concerned with the limitation of food intake, is termed the *satiety center.*

Lesions in a region neighboring the satiety center, the lateral hypothalamic area (Figure 7-1, "presumed feeding center") also result in a dramatic change in food intake: the animal stops eating completely, and characteristically starves to death without regaining its appetite. In this case a function has been lost, and so it is inferred that the destroyed area was responsible for the performance of the behavior which is no longer present. Since the lateral hypothalamic area is apparently involved in initiating eating, it is termed the *feeding center.*

Two centers are involved in the regulation of food intake, then, one apparently responsible for maintaining eating and the other for inhibiting eating. The existence of such pairs of opposing centers is quite common in the nervous system, and one might expect, arguing from analogy to what is known to be true for other areas of the brain, that there might be neural pathways between the feeding and satiety centers responsible for maintaining an inverse relation between the behavior of the two centers. More specifically, one might expect to find that as neurons in the feeding center increased their nerve impulse frequency, neurons in the satiety center would decrease their frequency, and conversely. Finally, these centers would, of course, alter their behavior in an appropriate fashion in response to some stimulus within the organism related to the state of the animal's nutrition. The concentration of blood sugar (glucose) has been frequently implicated in this role, and some workers in the field maintain that nerve impulse frequency in the appetite centers depends upon the blood sugar level: low levels causing decreased activity in the satiety center and increased activity in the feeding center.

On the basis of results of lesion studies, together with analogies to other systems, we have constructed a tentative model for the hypothalamic appetite centers (Figure 7-2). There are, of course, a number of alternative interpretations of the lesion studies, but the preceding model will serve to give some notion of the type of argument frequently made on the basis of brain lesions. Having constructed such a model, one may then make predictions about results expected from other types of experiments. For example, we should expect that stimulation of the feeding center would result in eating, while stimulation of the satiety center would cause a hungry animal to stop eating. Further, it is anticipated that the EEG recorded in the appetite centers might be different in hungry and satiated animals, and that the nerve impulse frequencies recorded from feeding center neurons would increase with hunger, while neuron impulse frequencies in the satiety center would decrease.

Electrical stimulation of the feeding center does indeed produce

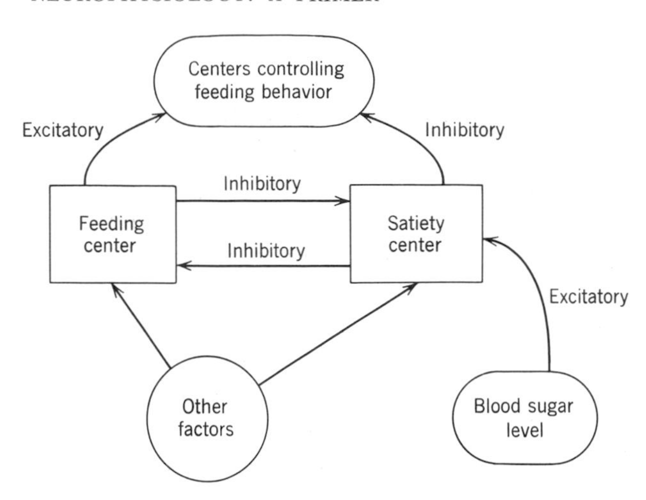

Figure 7-2. Model for the hypothalamic feeding centers. *Other factors* are such things as amount of food in the stomach and influences from centers controlling water intake.

eating in a fully satiated animal, while stimulation of the satiety center results in the inhibition of eating in a food deprived animal. Thus the results of stimulation experiments are in agreement with those of the lesion experiment: where removal of a region produces one effect, stimulation of that region produces the opposite effect. There is, moreover, one further bit of evidence from stimulation experiments which is in agreement with the model outlined above. According to the model, the feeding center neurons should be inhibited by the satiety center neurons when the animal is satiated. Since the feeding center neurons are inhibited in satiation, it should take increasingly larger stimuli to elicit eating as the animal becomes satiated. This is exactly what is observed in experiments: as the animal eats, an increasingly larger stimulus is required to maintain the eating.

Although it is difficult for technical reasons to compare appetite center EEGs in the deprived and satiated states, studies recording the EEG before and after the injection of glucose do indeed show changes which are located in the proper regions of the hypothalamus. Furthermore, extracellular recordings from neurons in these regions reveal that (presumed) satiety center neurons usually increase their nerve impulse frequency following the injection of glucose, while (presumed) feeding center neurons simultaneously decrease their frequencies. Finally, stimulation of the feeding center has been shown to inhibit neurons in the medial hypothalamus (region of the satiety center), while medial stimulation in turn inhibits neurons in the lateral hypothalamus (feeding center neurons), again in agreement with the tentative model.

Altogether, then, results of lesion, stimulation, and recording experiments are mutually consistent and in agreement with a model derived from the lesion studies and from experience with other, more thoroughly studied systems. It must be emphasized that the model itself must not be taken too seriously, since our knowledge in the field is still quite rudimentary; even the results of many of the experiments cited above are open to serious criticism. In spite of shortcomings, however, the model was presented for two reasons. First it illustrates the type of inferences one usually makes from the various techniques used, and second, it is, in general form, typical of the type of model most usually arrived at from studies of neurons in aggregate. No matter what function of the nervous system is under study, one is typically led to a picture involving several (or perhaps many) centers which are interrelated (often by inhibiting one another), and are either excitatory or inhibitory to a common center responsible for the final performance of the function in question. The extent to which these models are artifacts of the methods used must await further investigations; in the meanwhile, the lesion, gross recording, and stimulation techniques offer a way to obtain information which is otherwise not available.

Chapter 8

MEMORY

Although we have described properties of neurons which make it possible for the nervous system to carry out certain basic operations, no mention has been made of special neuron properties that might be used to perform one of the nervous system's most important and conspicuous functions, the storage of information. The question of the neuronal basis for memory, to which we now turn, brings us to one of the areas of neurophysiology where our knowledge is least advanced.

It is important, although often difficult, to distinguish between two categories of questions about memory. The first deals with the neural basis for the phenomena considered by the psychologist under the heading "learning and memory." Learning and memory at the behavioral level are very complex indeed, involving such things as the acquisition and organization of information, its storage, and problems of information retrieval. For example, one might ask why, in terms of neural mechanisms, some things are learned and others are not, or why early memories generally persist better than later ones. Understanding the neural mechanisms of memory at this level is not very different from understanding the neurophysiological basis of behavior altogether. Throughout the preceding chapters emphasis has been placed not on gross behavior of the nervous system, but rather on properties of single neurons. Consequently, one might ask: "What properties of the neuron make possible the storage of information?" Questions in this second category relate, then, to the problem of special neuron properties used for information storage as opposed to the neurophysiology of learning and memory in the broad sense. The following discussion will not be concerned with the explanation of learning and memory viewed as behavioral phenomena, but only with the more restricted issue of possible mechanisms for long term information storage in the nervous system. Since the neural events discussed in preceding chapters are measured on a time scale of milliseconds or hundreds of milliseconds, while memories may persist for years, we anticipate that some additional neuronal properties will underlie information storage in the nervous system. The question, then, is: "What are these additional properties?"

A number of theories have been put forward for the neural basis of memory, but none of them rest upon convincing evidence. In fact, it is unfortunately true that we do not even know precisely what facts a hypothetical memory mechanism must explain. Although we do know many things about memory as a behavioral phenomenon, very little is known about information storage at the cellular level. One might expect that properties of the cellular information storage mechanism could be deduced from the extensive store of data about memory in the psychological literature, but this is not generally the case. Because inferences about neuronal mechanisms are frequently drawn from behavioral data, it is worth illustrating the difficulties one encounters by reference to some specific examples. It is a common experience that learning often proceeds gradually, and that acquisition of a piece of new behavior requires considerable practice. From this, it is tempting to suppose that whatever neuronal changes underlie information storage also occur in a gradual, continuously graded manner. Common experience, however, together with experimental evidence, show that learning can occur very rapidly, and that a piece of behavior can be learned in only one trial with essentially no practice. Such experiments suggest that neurons involved in learning do not change gradually, but have only two states: "learned" and "not learned." Thus from the behavioral evidence, it is not clear whether neurons involved in storing information change gradually or whether they "learn" in an all-or-none manner (or both). Furthermore, either of the possibilities is compatible with the behavioral data. For example, gradual learning could simply be a reflection of the accumulation of increasing numbers of cells which had "learned" in an all-or-none way, whereas one-trial learning could be a switching from one type of behavior to another (learned earlier in a gradual manner) in a situation similar to others previously encountered.

The unfortunately prominent phenomenon of forgetting would seem to require an explanation in terms of neuronal properties, but again, it is not clear from the behavioral data precisely what is to be explained. Experiments with both humans and animals indicate that forgetting is not always (and perhaps not ever) a spontaneous fading of memories, but rather the learning of new responses which are incompatible with the old "forgotten" response; if this learning of competing responses is eliminated, there is no detectable forgetting. For example, pigeons that have learned the (for them) difficult task of telling the difference between two visual patterns, can still discriminate between the patterns perfectly after not having seen them for a considerable fraction of their lives, if they have no occasion to learn things which are at all similar. Even if there is no spontaneous forgetting, it is not clear what a theory of memory must explain with respect to the learning of competing responses. When new learning occurs, does the neuron involved in storing "forgotten" information unlearn, and return to its previous state, or is information storage on the cellular level a permanent, irreversible affair?

The preceding discussion should serve to illustrate the difficulty encountered in attempting to make inferences about the nature of the cellular information storage mechanism from behavioral data; one typically arrives at conflicting alternatives which can finally be decided between (or rejected completely) only by neurophysiological experimentation with a suitable preparation. We shall not pursue further this question of properties of the neuronal memory mechanism as inferred from behavioral information, but instead consider some other types of information about the neurophysiological basis of memory obtained from experiments combining behavioral observations with experimental manipulations of neuron function. By combining techniques for studying neurons in aggregate with those of experimental psychology, it has been possible to gather some more direct information about the nature of memory at the cellular level. From these studies it appears that memory is a composite phenomenon comprised of two different modes of information storage, each with distinct properties.

On the basis of both clinical experience with humans and animal experimentation two different types of memory, **recent** and **remote,** have been distinguished; these two types might equally as well have been termed **temporary** and **permanent.** Recent memory "contains" events occurring within about the last half hour or so and it is disrupted if normal brain function is interfered with (as by anesthesia), whereas remote memory may contain information stored over periods of years, and is not lost even if massive interruption of normal brain function occurs. In fact, remote memory is so resistant to disruption that procedures sufficiently drastic to cause death of large numbers of neurons (as asphyxia) apparently do not necessarily result in detectable memory deficit.

The existence of two types of memory is suggested by observations of brain damaged humans who have lost their remote memory (from the time of their brain damage onward) but who have recent memory intact. These people can remember recent events perfectly well, but by lunchtime have no idea of what they ate at breakfast. Nevertheless, they can remember events occurring before the time they sustained their brain damage in an apparently normal fashion. Evidently then, recent memory is intact, but the mechanism for permanent storage has been lost. Although such clinical observations are interesting and suggestive, analysis of the properties of recent and remote memory, (such as the length of time necessary for information stored temporarily to become a permanent memory) depends primarily on animal experiments. The following hypothetical experiment indicates how this type of research proceeds. Food is put in, for example, the right arm of a T-shaped maze, and hungry rats are placed in the maze and allowed to run to either the right or left arm, right turns being rewarded by food. The rats are divided into groups, and at some fixed time after each trial in the maze (the length of time depending on the group) they are anesthetized. The times between trial and anesthesia might be 10 seconds, 30 seconds, 1 minute, 5 minutes, 10 minutes, and so on up to 24 hours. After forty trials, the degree to which each group of rats learned to turn right would differ according to time between trial and anesthesia: rats that received anesthesia 10 seconds after each trial would do only a little better than chance (50% turn right), whereas rats in the 5-minute group would have learned about half as much as those in a control group (no anesthesia) who would nearly always turn right; animals in the 1-hour group and longer would show no effect of the anesthesia (see Figure 8-1). The experiment just described is not an actual experiment, but rather a hypothetical one illustrating the general technique and typical results. With different types of experiments somewhat different results are obtained. Nevertheless, providing that certain precautions are taken to avoid artifacts, the results are always qualitatively the same as those described.

Evidence from a number of different types of experiments indicates, then, that there are at least two types of memory mechanisms. One type, temporary or recent memory, depends on normal nervous system functioning. The second type, permanent

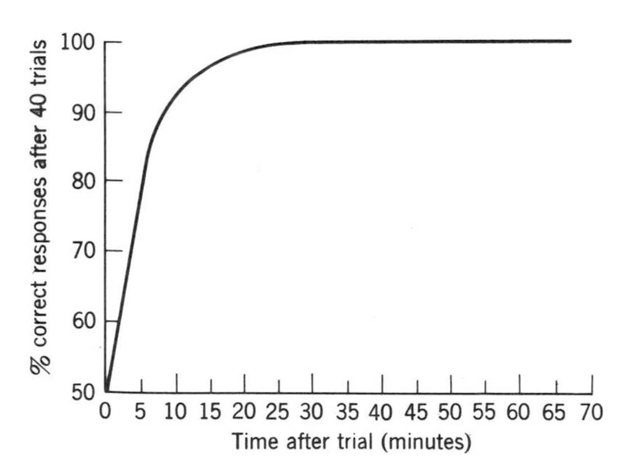

Figure 8-1. Results of an hypothetical experiment giving amount of learning (% correct responses for a group of rats) as a function of the time between the learning trial and onset of anesthesia.

or remote memory, does not depend upon normal nervous system function for its maintenance, possibly suggesting some biochemical or anatomical changes as underlying this type of information storage. Although not immediately translatable into inferences about the exact nature of the storage mechanism, such experiments do place quite definite requirements on any memory theory; furthermore they serve to emphasize the obvious point that no single, simple proposed mechanism can be expected to solve the problem of the cellular basis for memory.

Although the preceding discussion has been in no sense exhaustive, it does nevertheless adequately reflect the limited extent of our information about the cellular basis for information storage. The main reason for our extensive ignorance in this area is that no preparation is presently available which permits the study of learning and memory at the level of the individual cell; to date, no one has been able to observe a change in the behavior or properties of a neuron and say with assurance that the change was directly related to the memory mechanism. In fact, the situation is even somewhat worse than this. With one or possibly two exceptions, it is not known where in the brain specific memories are stored, or which parts of the brain do or do not have the capacity for information storage. Until we can locate cells responsible for storing information, it is unlikely that much progress can be made toward understanding the cellular memory mechanism.

In trying to locate neurons to use for studying learning at the cellular level, two different approaches are employed. Experimenters may attempt to locate cells involved in learning in simple organisms with a limited total number of neurons. As yet, such neurons have not been definitely identified, although they presumably exist since the organisms in question generally do exhibit learning. In the second approach, neurons involved in learning are sought in vertebrates, usually mammals. Here, despite the enormous numbers of cells present, some progress is apparently being made.

The best documented case of localization of memory involves a region of the brain known as cortical area 21. If monkeys are taught to distinguish between two slightly different patterns, and if area 21 then is removed (on both sides of the brain: the brain is symmetrical about a dividing plane, as are the face and head), the monkey can no longer remember the distinction he has previously learned, although vision is apparently normal; furthermore, this discrimination can be relearned, although with greater difficulty than before. If area 21 is removed before the animal is taught the discrimination, the process of learning to distinguish between two patterns is slowed down. Finally, there are changes in the electrical activity of area 21 which accompany the learning of the visual discrimination. These data, together with some others, make it reasonable to conclude that the information necessary for performing the discrimination is stored in area 21. This is not, however, the only possible interpretation; it is possible, for example, that no information is actually stored in area 21, but rather that this area is vital for retrieving information stored elsewhere, or for interpreting information already retrieved. Nevertheless, area 21 is the location best established as a memory site. Indeed, there are very few other brain regions the removal of which is known to disrupt memory at all.

One of the reasons why we have so little idea as to where in the brain memories are stored is that the same information is evidently stored in more than one location. If memories are stored in a number of sites, it is very difficult to discover their location since all the separate sites must, for example, be simultaneously removed to produce a memory deficit. The fact that memories are frequently stored in multiple locations is shown most clearly by experiments demonstrating that the same information is simultaneously placed in both sides of the brain.

Just as the body has a right and left side, so does the brain, each half being a mirror image of the other. Therefore, each nucleus or fiber tract in one half of the brain has a companion nucleus and fiber tract on the opposite side. Communication between the two halves of the brain is accomplished through large fiber tracts; the largest of these in the higher mammals is known as the **corpus callosum.** To a certain extent information from receptors on one side of the body is sent initially to only one of the two halves of the brain. For example, information about pressure on the left side of the body is transmitted directly to the right side of the brain (fiber tracts transmitting information about pressure cross from one side to the other before finally synapsing). In the case of the eye, the situation is somewhat complicated by the fact that each eye sends information to both halves of the brain: part of the visual information from the right eye is transmitted to neurons in the right half of the brain, while the remainder of the right eye's information travels along a fiber tract which crosses to terminate on neurons in the left half of the brain. The point of crossing is known as the **optic chiasm.**

Suppose that an experimental animal, such as a cat or a monkey, is taught to press one key if shown a square and press a second key if shown a circle, and further suppose that the response is learned with the animal's right eye blindfolded. If the blindfold is switched from the right to the left eye, it should surprise no one that the animal can still perform the required response correctly. A more surprising result is seen in the following situation: suppose that the crossing optic tract axons are cut before the animal is taught to discriminate between the circle and the square with the right eye blindfolded. In this case, no information from the left eye can reach the left side of the brain directly. Yet, when the blindfold is switched from the right to the left eye, the animal can still perform the discrimination despite the fact that visual information is now going first to a side of the brain which has had no experience with the problem. Obviously, the two halves of the brain are exchanging information. This exchange of information can occur only over the fiber tracts connecting the two halves, the main one of which, it will be recalled, is the corpus callosum.

We are now in a position to investigate whether or not information is stored on both sides, or on just one side of the brain. If memories necessary for performing the discrimination are stored on both sides of the brain, an animal (in which crossing optic nerve fibers have been cut) trained with the right eye blindfolded should be able to perform the learned response equally well with the left eye blindfolded even after the corpus callosum is cut. On the other hand, if memories are stored only on the one side, cutting the corpus callosum should make these memories inaccessible when the blindfold is switched (since the tracts over which information travels between halves of the brain are no longer available, and there would be no way for the right side of the brain to communicate with the storage sites in the left half). Experiments have shown that animals taught a discrimination with one eye can still perform the response using the opposite eye even after the corpus callosum has been cut; information must then be transferred from one side of the brain to the other and stored on both sides.

If information arriving on one side of the brain is transferred to the other side via the corpus callosum for storage, it should be true that animals taught a response with the left eye should not be able to perform the response using the right eye when the crossing optic nerve axons and the corpus callosum have both been destroyed prior to learning the response. Experiments show this to be true. Having learned a response with the right eye blindfolded, an animal shows no sign of ever having known the response when the blindfold is switched to the left eye, although the same response (or a different one) can also be taught to the right side of the brain. These experiments demonstrate, then, that the halves of the brain can be made to function quite independently.

Taken together, the observations outlined in the preceding paragraphs indicate that the same information may be stored simultaneously in both halves of the brain. Thus a particular memory is, at least in certain situations, stored in two (or more) separate locations. Indeed, other experiments have indicated that information relating to a particular learned response is probably stored simultaneously in many separate locations. Since one of the best methods of locating cells involved in information storage is to remove a brain area and study the memory deficit produced, the fact that memories are stored simultaneously in multiple sites makes the situation more difficult rather than easier, as might be expected. In order to produce a memory deficit, it is necessary, as was pointed out earlier, to remove all of the locations in which that particular memory is stored. If there are a number of different locations, it is naturally quite difficult to remove by chance all of the locations in which the memory is stored. Thus it is difficult to produce a memory deficit at all without destroying major portions of both sides of the brain, in which case one knows little more than that the memories were somewhere in the brain. Although progress has been slow, continued work together with the use of some newer techniques (such as the brain "splitting" method just described) should eventually provide information on where various types of memories are stored, and thereby provide better preparations for studying learning and memory at the single neuron level.

The situation regarding the cellular basis for memory can be summarized briefly as follows: from behavioral and physiological experiments we know a few general properties of memory at the cellular level such as the fact that the neuronal changes underlying memory are long lasting and very resistant to gross disruption of normal brain function, however, we do not know many properties of memory which are prerequisite for an understanding of the information-storing mechanisms of the nervous system, for example, whether learning at the level of the single cell is all-or-none or gradual. It is known that essentially the same information is stored at multiple sites, and one or two of the storage sites have probably been discovered, but exactly which parts of the nervous system are necessary for memory is not known. It is not even definitely known that information is stored in neurons; it has been suggested that information storage occurs in glia (see page 8, Chapter 1), although there is no evidence to substantiate this theory. As was noted previously, the primary problem is that a preparation suitable for studying learning and memory at the cellular level is not presently available, and rapid progress in the field cannot be expected before such a preparation is discovered. Considerable effort is currently being devoted to a search for a suitable preparation, and experimenters interested in memory are exploring alternative approaches. Perhaps the most actively pursued of these alternatives is the study of physiological and biochemical memory models, that is, biochemical or neuronal systems which exhibit behavior in some way reminiscent of memory.

Since most changes associated with neurons are exceedingly shortlived compared to the long-lasting effects of memory, a number of workers are involved in looking for long-lasting changes in the behavior of neurons in the expectation that such changes are related to memory. Some of the long-lasting changes which have been discovered provide plausible information storage models. For example, several workers have found neurons that would alter their average impulse frequency apparently indefinitely (at least for the duration of the experiment) in a reversible way following appropriate stimulation lasting a number of minutes. The effect has not been very extensively investigated as yet, and has no demonstrated connection with learning. Nevertheless, the neurophysiologist at once wonders what as yet unstudied properties of some neurons could explain such behavior, and is tempted by the possibility that these properties may serve as the neuronal substrate for recent memory.

In addition to the neuronal models just mentioned, there has been much recent attention to biochemical memory models. Several systems studied by the biochemist have properties in some way reminiscent of memory and learning, and the hope is that nature has been sufficiently economical to evolve only one or a few memory-like mechanisms. To date, the idea that information storage in the nervous system is in some way connected with RNA has been the most popular, but more recently interest has turned to enzyme induction and antibody formation as attractive model systems.

Since RNA has been termed a "memory" molecule (or at least an information-transferring molecule), it has been argued that memory in the nervous system might be stored in RNA changes; in fact, neuronal RNA changes in "learning situations" have been demonstrated. To the extent that a neuron that has learned is different from what it was before it learned, it is not unreasonable that it should manufacture somewhat different proteins, or perhaps different proportions of the same proteins. With sufficiently refined techniques, concomitant RNA changes could certainly be detected. One would indeed expect, therefore, to find RNA differences correlated with learning in a neuron, if for no other reason than that the number of nerve impulses per unit time would probably be altered (and changes in impulse frequency have been demonstrated to have biochemical concomitants). Granting, then, that RNA would be expected to change in some way during learning, the mechanism of information storage does not become any clearer. Conceivably, it might be helpful to know just how RNA changed, but the methodological difficulties here are enormous; since we do not even know which neurons are responsible for information storage, it becomes difficult to separate RNA changes involved in learning itself from nonspecific changes related to the fact that the nervous system is behaving differently during and after learning than it did before.

There are analogies between the phenomena of enzyme induction, antibody formation, and learning, and it is possible that a full understanding of the biochemistry of enzyme induction (or the more general problem of control in biochemical systems) and of antibody formation may someday have relevance to the problems of information storage in the nervous system; at present the relevance to neuron function is not clear.

We are just now, it would appear, on the threshold of making major advances toward discovering properties of neurons which serve as the basis for information storage in the nervous system. At present, however, and probably for the foreseeable future, understanding the neuronal mechanism underlying learning and memory may be expected to constitute one of neurophysiology's central problems.

Chapter 9

MECHANISMS OF NEURAL ACTIVITY: QUALITATIVE TREATMENT

The approach of the preceding chapters has been to first describe a number of basic properties of neurons, and then investigate some implications these properties have for nervous system functioning. The properties of membranes which result in action potentials and PSPs have been accepted as given, and nervous system functions have been explained in terms of these basic concepts. Thus far however, we have not inquired into the physical or physico-chemical mechanisms underlying the action potential and the PSP. Chapters 9 and 10 will present the current view of the mechanisms responsible for those properties which have been simply described before; thus we shall proceed one level deeper into the explanation of nervous system function. A discussion of the mechanisms responsible for the nerve impulse and PSP has been deferred until now because its understanding requires more prerequisite knowledge than material covered previously; in fact, more than just elementary physics, chemistry, and mathematics are needed for a full understanding of this subject matter. To make the presentation more accessible this chapter will present a qualitative version of the ionic basis for nervous system function, whereas the following chapter will give a more accurate quantitative treatment for readers with the required background.

Heretofore the behavior of nerve cells has, for simplicity, been described entirely in terms of voltage changes. For instance, *action potential* and *PSP* have been identified with their characteristic patterns of membrane potential variations; *passive spread* has been defined in terms of the effect of a membrane potential change at one point on the membrane potential at other points of the neuron. However, when the purely descriptive approach of the preceding chapters is abandoned in favor of a consideration of mechanisms underlying neuron behavior, it becomes necessary to take variables other than voltage into account. Particularly important, in addition to membrane potential, are **membrane current** (the movement of ions across the membrane) and **membrane permeability** (how easily a particular kind of ion penetrates the neuron membrane). To illustrate the various factors which come into play in the generation of the action potential, we shall summarize, at first without explanation, the events believed to constitute the nerve impulse.

THE RESTING POTENTIAL AND THE ACTION POTENTIAL

By an incompletely understood mechanism known as the **sodium pump,** sodium ions are primarily kept outside of neurons, whereas potassium ions are primarily kept inside; that is, the potassium ion concentration is high within neurons and low in their surrounding fluids, and sodium ions are in high concentration outside of neurons and low within. In the resting state, the neuron's membrane is much more permeable to potassium than to sodium, a fact used later to explain the resting potential.

The action potential itself is, as we shall see in more detail later, a consequence of a sequence of alterations in the membrane permeability to sodium and potassium ions (Figure 9-1). Sodium permeability is very low in the resting state, increasing dramatically during the rising phase of the action potential, and then returning to a low value on the declining phase of the action potential. Although potassium permeability is also low in the resting state, it is nonetheless much greater than sodium per-

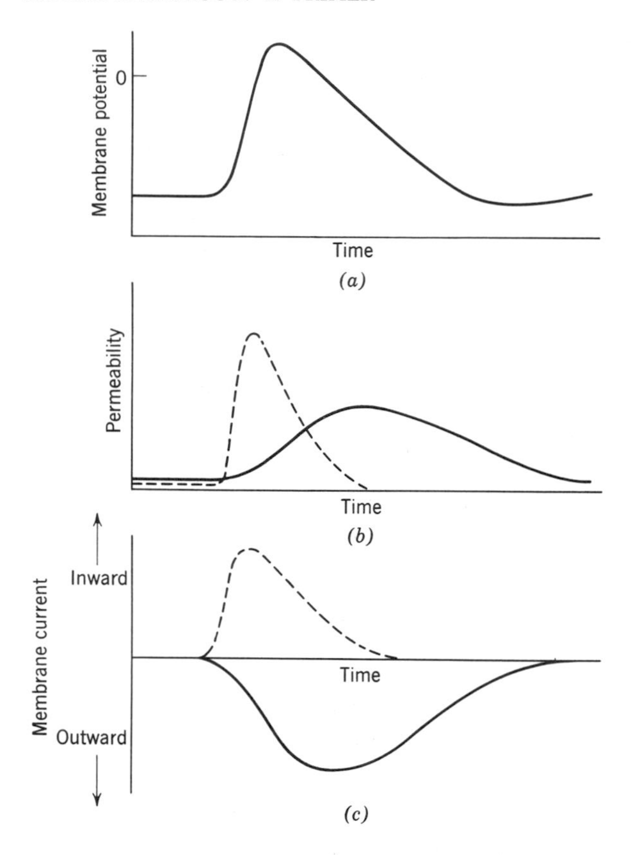

Figure 9-1. Events occurring during the nerve impulse. (*a*) Membrane potential fluctuations constituting the action potential. (*b*) Permeability to sodium (dotted line) and potassium (solid line) coincident with membrane potential changes shown in (*a*). (*c*) Membrane current during the action potential. Sodium current is represented by the dotted curve and potassium by the solid curve.

meability. The potassium permeability increases gradually during the rising phase of the action potential, reaches a maximum during the decline of the action potential, and finally returns to its resting level after the electrical signs of the nerve impulse are nearly over. Associated with these changes in permeability is a sequence of ionic current flows also illustrated in Figure 9-1. Sodium flows into the neuron during the depolarization phase of the action potential, and potassium flows out during the return to the resting potential. In the more detailed consideration of the relationship between ionic permeability, ionic current, and membrane potential which follows, it will be seen that the basic events underlying the action potential are the alterations in sodium and potassium permeability of the membrane. A detailed understanding of the physical (or molecular or physico-chemical) mechanism responsible for these permeability changes still constitutes one of neurophysiology's central problems.

From the preceding summary it can be seen that the factors involved in the production of the action potential are membrane permeability (to specific ions), membrane current, membrane potential, and the concentration differences of sodium and potassium maintained across the membrane by the sodium pump. We must now investigate the relationship between these various factors.

We know from physics that a voltage difference between two regions results in the movement of ions: positively charged ions tend to move from the more to the less positive region, whereas negatively charged ions move in the opposite direction. In the resting neuron, for instance, positive ions tend to move from the outside to the inside, and at the same time negative ions tend to pass from inside the cell to the surrounding fluids (since the inside is about 60 millivolts negative relative to the outside). Thus the membrane potential (the inside-outside potential difference) may be thought of as a driving force causing ionic current to flow through the membrane. The actual magnitude of the ionic current depends, of course, on the nature of the neuron membrane: if the membrane happens to be very permeable to a particular ionic species, the current will be large, and if the membrane is relatively impermeable, the current will be small.

Rather than speak of *permeability* as we have in the preceding paragraphs, it is customary to specify the ease with which a type of ion passes through the membrane by the magnitude of the **conductance** for that ion. A large sodium conductance, for example, means that a relatively large number of sodium ions will flow through the membrane per second given a (perhaps) 50 millivolt membrane potential, whereas a small chloride conductance would mean that the same 50 millivolt voltage difference would cause relatively few chloride ions to traverse the membrane. The membrane conductance for an ion may be given more concrete meaning if thought of as the inverse of the frictional resistance offered to that ion by the membrane.

In addition to the voltage difference across the neuron membrane, concentration differences constitute the second important force driving ions through the membrane. Just as a positively charged ion moves from a more to a less positive region, so does an ion move from a region of high concentration into a region of lower concentration; this phenomenon is known as **diffusion.** Since sodium has a higher concentration on the outside of the membrane, it tends to diffuse into the neuron, while potassium moves in the opposite direction. The number of ions moving through the membrane per second depends on the size of the concentration difference across the membrane, and also on the conductance of the membrane to the ion in question. For example, if sodium conductance is low, sodium ions will diffuse slowly from outside to inside, and if sodium conductance is large, they will flow in more rapidly.

Altogether, then, three factors jointly determine the flow of a particular ion across the membrane: membrane potential, concentration differences, and conductance. The interaction of these factors is illustrated in the following model situation, which is a sufficiently accurate representation of the actual neuron for present purposes. Suppose that the concentration of potassium on the inside of a neuron is ten times that of the outside concentration and the outside concentration of sodium is ten times that on the inside; the negative ion balancing the positive sodium and potassium ions is chloride so that the model neuron has no net charge. These concentration gradients are maintained unchanged by the sodium pump even though sodium and potassium are moving across the membrane by diffusion. Having fixed the concentration differences, we must next make some assumptions about the conductances for our model. We shall suppose the chloride conductance to be zero (the membrane is completely impermeable to chloride), the sodium conductance to be relatively low, and the potassium conductance to be relatively high; this represents the resting state of the neuron. Finally, the *initial*

voltage difference across the membrane will be assumed, for the purpose of illustration, to be zero.

Since we have assumed that initially there is no voltage difference across the membrane to act on the ions, in the first instant of time they will be driven through the membrane entirely by concentration differences. By assumption, the potassium conductance is higher than the sodium conductance, and the potassium and sodium concentration differences are equal (although in opposite directions); therefore more potassium than sodium ions will pass through the membrane in the first instant of time. Since small amounts of sodium will move into the "neuron," and larger amounts of potassium will move to the outside (and chloride will not move at all), the result will be a slight excess of positively charged ions on the outside of the membrane. This excess of positively charged ions will cause a voltage difference across the membrane, the inside becoming negative relative to the outside.

The membrane potential which results from any given charge difference depends on the physical characteristics of the membrane, since membrane potential (inside-outside voltage difference) is a measure of the amount of work necessary to bring about the particular charge difference. Factors such as thickness of the membrane determine the amount of work necessary to create a charge difference, and are therefore reflected in a membrane potential resulting from that charge difference. In fact, the membrane potential is proportional to charge difference, the proportionality constant being termed **membrane capacitance.** Thus the various physical attributes of the membrane which determine what membrane potential will result from a given charge difference are contained in a number, **membrane capacitance,** which characterizes the neuron membrane. Capacitance is defined in such a way that a large membrane capacitance means that a large charge difference will cause a relatively small membrane voltage. Capacitance may therefore be thought of as the capacity a membrane has for accommodating a charge difference for a fixed voltage.

After the first instant of time the flow of ions will be determined not only by the concentration differences, but also by the voltage difference that has developed as the result of the previous instant's currents. In the second instant the number of potassium ions flowing out will be decreased because the concentration difference will be opposed by a voltage difference; the sodium current will increase, since the membrane potential as well as the sodium concentration difference is driving sodium ions from outside to inside. In the second instant of time, the number of potassium ions flowing out will probably again exceed the number of sodium ions flowing in, so that the outside will become still more positive than the inside.

This same process will occur in each instant of time as summarized in Figure 9-2, with the voltage difference across the membrane progressively building up, and tending to equalize the sodium and potassium flows. After a period of time, the volt-

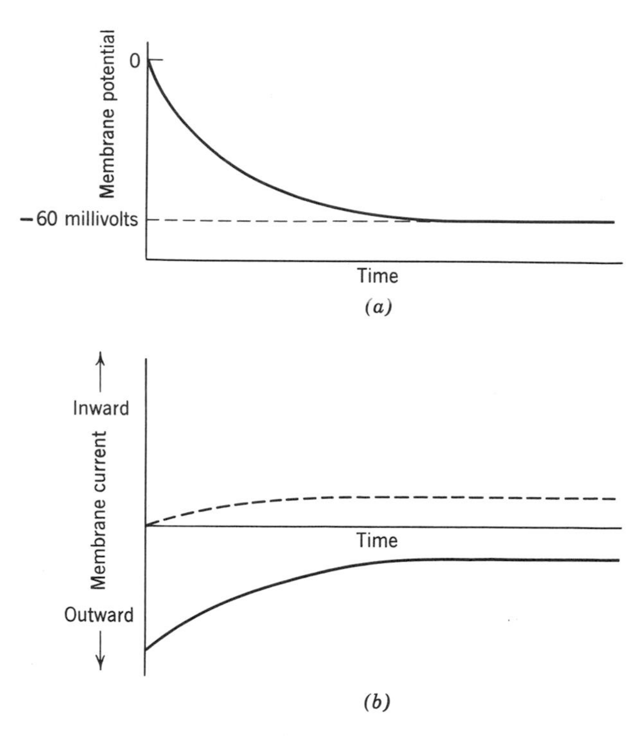

Figure 9-2. Approach to the "resting potential" in a model neuron. (*a*) Membrane potential initially at zero potential. (*b*) Sodium (dotted line) and potassium (solid line) currents associated with membrane potential in (*a*).

age difference will be sufficient to make the potassium and sodium currents equal. When this occurs, the number of charges moving to the outside will be the same as the number moving to the inside, and the voltage will no longer change (since the charge difference across the membrane will no longer change). What will have happened is that the excess of positive charges on the outside will have caused a membrane potential large enough to counterbalance the net outward flow of positive ions resulting from the inside-outside concentration differences and conductance differences.

In the process described, the movement of ions across the membrane resulted in charge differences causing a voltage difference which itself affects the flow of ions. One would also expect the movement of ions to change the concentration differences across the membrane, and thereby also affect ionic flows. In the neuron, and in our hypothetical situation, significant changes in concentrations do not occur as the result of ionic movement for two reasons. First, the number of ions moving across the membrane in a short period of time, several minutes, for instance, is generally small compared to the total number of ions within and surrounding the neuron, and insufficient to change the concentrations by an appreciable amount. Second, the sodium pump is constantly working to maintain constant concentration differences and manages, on the average, to move sodium out as rapidly as it is coming in. For these reasons, the concentration differences across the membrane may be considered to be constant over time in spite of the continual ionic flows through the neuron membrane.

The rate at which the steady-state membrane potential would be achieved in the preceding example depends jointly on the magnitudes of the membrane capacitance and of the total conductance (sum of sodium and potassium conductances). If the conductances are large, great numbers of ions will flow across the membrane each second, and the charge difference (and thus membrane potential) will build up rapidly. Conversely, if the conductances are very small, ions will cross at a low rate, and the voltage will change more slowly. Membrane capacitance is the number relating voltage difference to charge difference across the membrane; the larger the capacitance the greater the charge difference necessary to produce a given membrane potential. Therefore a large membrane capacitance has an effect opposite to that of a high conductance on the rate at which the final steady membrane potential is reached. If membrane capacitance is large, a relatively large number of ions must cross to produce (perhaps) a 1 millivolt change in membrane potential and the final membrane potential is reached more slowly. Since the total conductance and membrane capacitance have inverse effects on the rate of achievement of the steady-state membrane potential, their ratio, known as the **membrane time constant,** is a convenient measure of this rate.

It will be recalled from the preliminary description on page 115 that the neuron maintains approximately the sodium and potassium distributions assumed in the preceding example, and the potassium conductance considerably exceeds the sodium conductance when the neuron is in the resting state. By the process just described, the membrane potential arrives and is maintained at about 60 millivolts (outside positive). Thus the sodium pump, which maintains the concentration differences, together with the resting values of the conductances adequately account for the resting potential. It will also be recalled that the process underlying the generation of the action potential is a sequence of alterations in the sodium and potassium conductances. To understand how such conductance changes can account for the action potential, we shall first consider the effect of varying the conductances on our model neuron, and then investigate in more detail the role played by the inside-outside concentration differences.

We suppose, as before, that the sodium and potassium concentration differences across the model neuron's membrane are both about 10 to 1, with the high potassium concentration on the inside and an equally high sodium concentration on the outside. Initially we shall assume the potassium conductance, as before, to be considerably higher than the sodium, and we shall let the membrane potential start at its resting level (that is, at the steady-state level determined by the concentration and conductance differences). Suppose now that suddenly (as shown in Figure 9-3)

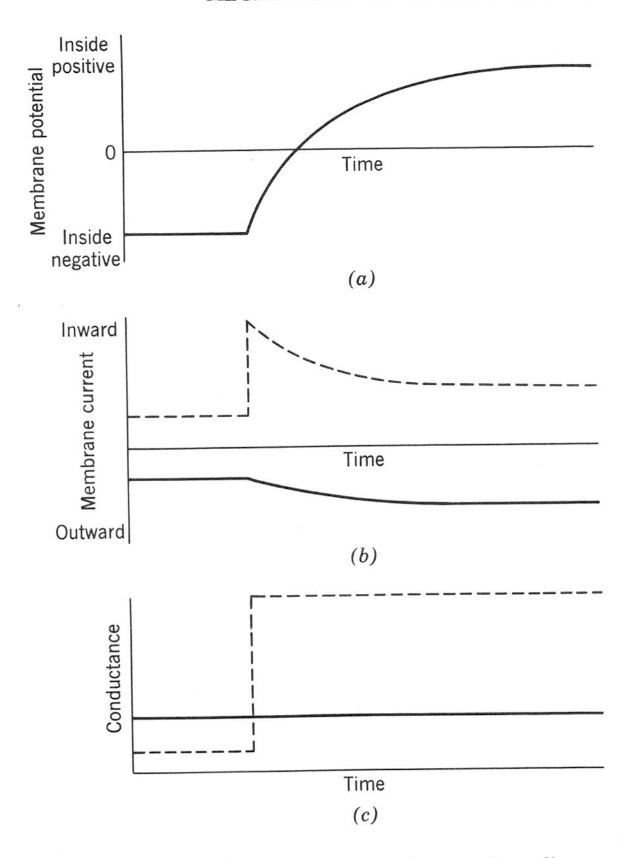

Figure 9-3. Response of model neuron to a step increase in sodium conductance. (*a*) Membrane potential change resulting from conductance change. (*b*) Sodium (dotted line) and potassium (solid line) currents associated with membrane potential in (*a*). (*c*) Conductance as a function of time resulting in curves (*a*) and (*b*). Sodium conductance is indicated by the dotted line and potassium conductance by the solid line.

the sodium conductance were to become larger than the potassium conductance. When this happens, the diffusional flow of sodium becomes much larger than that of potassium, resulting in the net flow of positive charges from outside to inside. Because the neuron is positive on the outside relative to the inside, the membrane potential also causes positive charges to move from the outside to the inside. Altogether then, there will be a rapid net movement of positive charges (sodium and potassium) from the outside to the inside. As this happens, the initial charge difference across the membrane diminishes (the outside will become less positive) and finally reverses. As the inside of the neuron becomes positive, the membrane potential moves positive charges *out* of the neuron, counteracting the inward flow caused by the concentration and conductance differences. Finally, by the same arguments employed on page 120 to explain the development of the resting potential, a new steady-state membrane potential will be achieved, with the neuron now positive rather than negative on the inside.

In the initial example, the model situation corresponding to the resting neuron, potassium conductance was much larger than sodium conductance, whereas the reverse was true in the second example; the higher potassium conductance resulted in a *negative* membrane potential (outside positive relative to the inside) while the larger sodium conductance caused a *positive* membrane potential. By essentially identical arguments, it can be seen that intermediate values of conductances will result in intermediate membrane potentials. For example, if the sodium and potassium conductances were exactly equal, then the diffusional flows of potassium and sodium ions would be equal, and a zero membrane potential would be required to have no net accumulation of charge on one side of the membrane. The steady membrane potential caused by any ratio of conductances is illustrated in Figure 9-4, where potassium conductance is held fixed and sodium conductance is varied from zero to many times the value of the potassium conductance. When sodium conductance is zero (only potassium can pass through the membrane), the membrane potential is maximally negative; as the sodium conductance increases, the membrane potential becomes less negative, passes through zero when the conductances are equal, and finally approaches a maximally positive value.

Having described the effect changes in the sodium-potassium conductance ratio have on the steady membrane potential, we must now investigate the role played by the concentration differences. Until now, we have assumed both sodium and potassium to be in a 10 to 1 ratio of concentration between the inside and the outside. How would results change if the inside-outside potassium concentration ratio were 20 to 1 instead of 10 to 1? Because

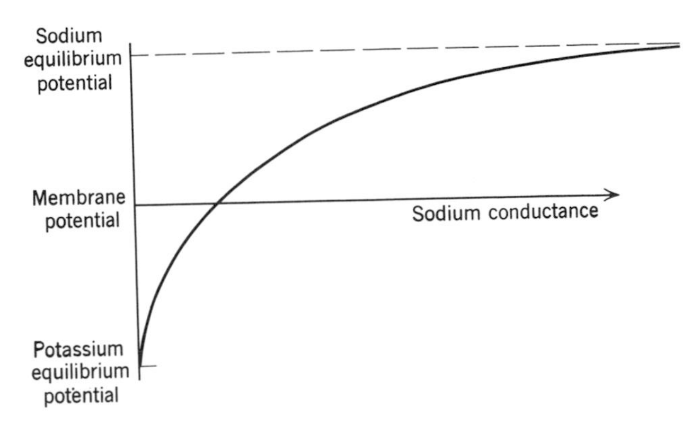

Figure 9-4. Steady-state membrane potential seen with various sodium conductances and a fixed potassium conductance. At the left of the figure sodium conductance is zero, and increases to many times the potassium conductance at the right.

the concentration difference is larger, the ionic flows due to diffusion will be greater, and a larger membrane potential will be required to counterbalance the ionic movements across the membrane caused by diffusion. Therefore, if potassium conductance dominates, that is, if the sodium conductance is so small that the sodium flow has very little effect, the steady membrane potential will be more negative than in the case where the potassium concentration ratio was 10 to 1. The maximal negative membrane potential (the one approached as the sodium conductance becomes negligible compared to the potassium conductance) thus depends on the potassium inside-outside concentration ratio; this maximally negative membrane potential is known as the **potassium equilibrium potential.** Similarly, the maximally positive membrane potential, approached as the potassium conductance becomes very small compared to the sodium conductance, is termed the **sodium equilibrium potential.** The effect of inside-outside concentration ratio on the equilibrium potential is illustrated in Figure 9-5.

Having described the relationship between concentration differences, membrane potential and ionic conductances, we now return to the events underlying the action potential. It will be recalled that on page 115 we described a sequence of conductance changes together with associated membrane potential fluctuations

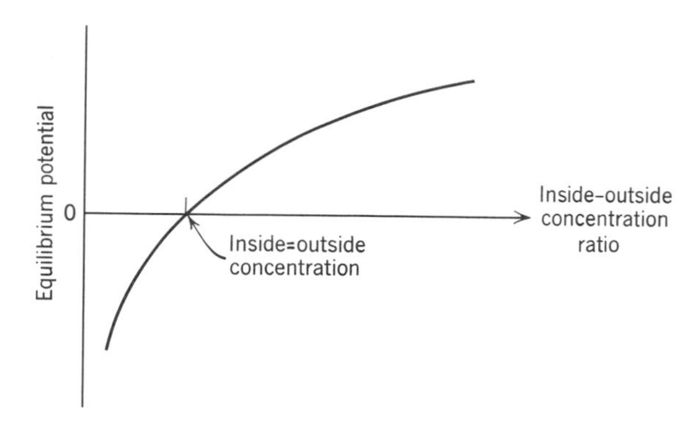

Figure 9-5. Equilibrium potential for an ion as a function of the inside-outside concentration ratio for that ion. If the inside and outside concentrations are the same, the equilibrium potential is zero.

and ionic current flows together constituting the nerve impulse. We are now in a position to understand how the conductance changes are basic to the production of a nerve impulse and why the membrane potential changes (the action potential) are a consequence of them. A discussion of why the conductance changes themselves occur will be presented shortly.

In the resting state, it will be recalled, potassium conductance is low and the sodium conductance is even lower. This leads to a more rapid movement of potassium than sodium, and thus to the accumulation of a net positive charge on the outside of the membrane resulting in the production of the resting potential. The first event in a nerve impulse is a rapid increase in sodium conductance which soon greatly surpasses the potassium conductance. Sodium then flows into the neuron faster than potassium is moving out, positive ions accumulate on the inside, and the membrane is depolarized. The greater the increase in sodium conductance, the faster the sodium flows in and the greater the depolarization (up to the sodium equilibrium potential). As the sodium conductance increases, the membrane potential approaches the peak of the action potential. This up-swing, in turn, causes the potassium conductance to increase more rapidly. By the peak of the action potential, the sodium conductance has started to decrease whereas the potassium conductance continues to increase. When the potassium conductance is relatively greater than the declining sodium conductance the outward flow of potassium becomes greater than the inward flow of sodium, and the membrane potential is driven back towards the resting (inside negative) level. Finally, as the resting potential is approached, potassium conductance decreases to its normal resting level, and the neuron is returned to its quiescent state.

We now turn to the question of how the conductance changes underlying the action potential occur. By means of experiments described in Chapter 10, Hodgkin and Huxley were able to demonstrate that both sodium and potassium conductance depend on membrane potential itself: the greater the depolarization, the greater the sodium and potassium conductance. This simple relationship between membrane potential and conductance does not fully describe the process, however. It takes time for the final conductance to be reached once the neuron has been depolarized for both sodium and potassium conductances. That is, a decrease in membrane potential does not cause the sodium and potassium conductances to increase instantaneously, but rather results in a smooth rise to the final value. Furthermore, sodium and potassium conductances do not increase at the same rate for the same amount of depolarization: sodium conductance generally increases more rapidly than potassium conductance. That is why an action potential can occur when the neuron is sufficiently depolarized. Once an EPSP depolarizes a neuron, for example, both the sodium and potassium conductances start to increase. However, the sodium conductance increases more rapidly, causing the neuron to be further depolarized, and thus causing further rapid increases in sodium conductance. A runaway situation develops, since each increase in sodium conductance results in a further depolarization, which in turn causes a new increase in sodium conductance.

The fact that sodium conductance increases more rapidly than potassium conductance thus explains the rising phase of the action potential. How is the return to the resting potential to be explained? Hodgkin and Huxley found that after a depolarization had first caused the sodium conductance to increase, it then caused it to decrease. That is, if a neuron is in some way depolarized and not allowed to return to the resting potential, the sodium conductance will first increase and then decrease in spite of the fact that the neuron is still depolarized. This phenomenon is known as **sodium inactivation.** The return of the action potential to the resting level is the joint result of sodium inactivation and increased potassium conductance. Once the sodium conductance has increased to a large degree, it again decreases (because of inactivation) although the neuron is still depolarized. Since the neuron is depolarized, the potassium conductance has increased by a large amount. Just as potassium conductance increases comparatively slowly when the neuron is depolarized, so does it again decrease fairly slowly once the neuron is no longer depolarized by a large amount. Thus the "turning off" of sodium added to the slow decline of potassium conductance account for the return of the action potential to the resting potential.

We have now seen, in a sketchy fashion, how the neuron's membrane potential is controlled by the sodium and potassium conductances, and how these conductances in turn depend on the membrane potential. The next step in an explanation of the nerve impulse would seem to be an elucidation of why the sodium and potassium conductances change the way they do in response to membrane potential alterations. This, unfortunately, is not yet understood, and constitutes one of neurophysiology's central problems. Indeed a considerable amount of research is at present being directed toward answering the question of exactly how the neuron membrane can selectively change its permeability to sodium and potassium ions in response to membrane potential changes.

MECHANISM OF POSTSYNAPTIC POTENTIALS

In Chapter 3 a distinction was drawn between electrically and chemically excitable membranes. An electrically excitable membrane, it will be recalled, produces its typical response (the action potential) when it is depolarized; in the terminology of this chapter, electrical excitability is a consequence of the fact that sodium and potassium conductances depend on membrane potential and can at the same time themselves produce membrane potential changes. Thus depolarization of a neuron membrane can result in an increased sodium conductance which in turn produces further depolarizations and so on. In contradistinction to this electrically excitable membrane, the ionic conductances in a chemically excitable membrane do not depend on membrane potential; if a chemically excitable membrane is depolarized, the sodium and potassium conductances do not increase but rather remain unchanged.

Although conductances in a chemically excitable membrane do not change in response to depolarization, they do change upon the application of an appropriate chemical. When a transmitter diffuses from an axon terminal to a postsynaptic membrane, conductance changes in that membrane occur and in turn produce membrane potential changes identified as the PSP. The size of the conductance change, and thus the size of the PSP, depends on the transmitter concentration. The specific nature of the conductance changes—whether of sodium, potassium, or some other ion—depends on the characteristics of the membrane and of the chemical structure of the transmitter substance. At an IPSP producing synapse, for example, the transmitter apparently acts to increase the potassium and chloride conductances (leaving the sodium conductance unchanged), thereby producing a hyperpolarization. The movement of positively charged potassium out of the neuron together with the inward movement of negatively charged chloride ions (in the real neuron, chloride is in higher concentration on the outside, with organic ions serving as the negatively charged intracellular ions) result in an increased net negative charge inside the neuron and a membrane hyperpolarization. The sequence of events during an IPSP is illustrated in Figure 9-6. When an EPSP is produced, sodium, potassium, and possibly chloride conductances increase. Here the situation is somewhat more complicated since potassium and chloride movement tends to hyperpolarize the cell, whereas sodium influx tends to depolarize it. However, the sodium conductance increase is much larger than the potassium and chloride, and the net result of the transmitter action is an inward movement of positive charges and the depolarization constituting the EPSP.

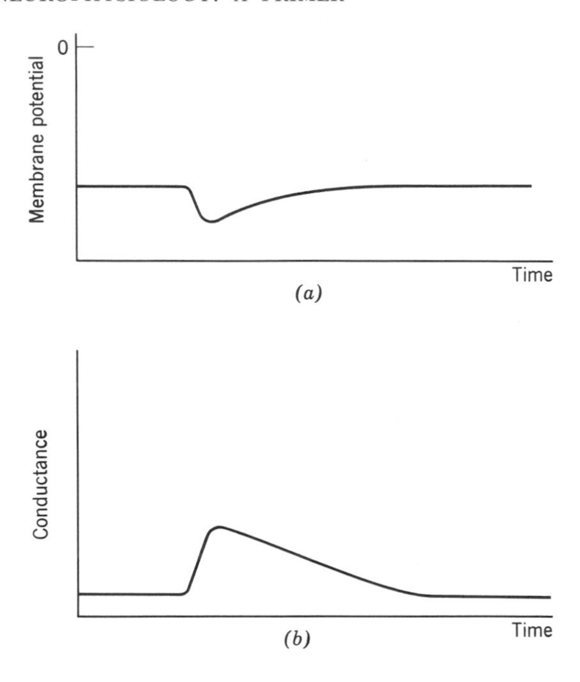

Figure 9-6. Membrane potential and conductance during an IPSP. (*a*) Membrane potential fluctuation constituting the IPSP. (*b*) Potassium and chloride conductance increase resulting in the membrane potential seen in (*a*).

The distinction between chemically and electrically excitable membranes is not, it will be recalled, always a sharp one, for some neuron membranes may be both electrically and chemically excitable. The important point in making this distinction is that two fundamentally different mechanisms exist in the nervous system for producing conductance changes. In some cases, conductance changes are caused by membrane potential changes and in other cases result from the action of a chemical transmitter. These two mechanisms have very different properties. If conductance depends solely on membrane potential, conductance changes tend to produce further conductance changes because of the relationship between conductance and membrane potential. In contrast, if conductance depends primarily on transmitter concentration, conductance changes, and also membrane potential changes, merely reflect the transmitter concentration and there is no influence of conductance back onto itself. The first situation, where conductances depend on membrane potential, gives rise to the threshold phenomena, the all-or-none type of response and other properties discussed in connection with the physiology of axons in Chapter 2; in the cases where conductances depend on transmitter concentration, properties such as temporal summation described in Chapter 3 in connection with the physiology of dendrites are observed.

In Chapter 3 and in the preceding discussion we have said that during the PSP the membrane potential reflects transmitter concentration. From this, one might infer that the time course of transmitter concentration, of the conductance changes, and of the membrane potential are all exactly parallel. This is not always true, however. If transmitter is destroyed relatively slowly, so that its duration of action is perhaps several hundred milliseconds, the time course of transmitter concentration is well reflected in the form of the PSP. In some instances, however, the transmitter molecule is so rapidly inactivated that the conductance change produced by the arrival of a single nerve impulse lasts only one or several milliseconds. In this situation the membrane potential change is considerably longer than the several milliseconds over which transmitter acts. The reason for this lack of coincidence between conductance and membrane potential changes has already been illustrated on page 123 (see Figure 9-3). Because the processes of ion transfer across the membrane take time, conductance changes and membrane potential changes are not precisely parallel, but rather membrane potential follows conductance rather sluggishly. Thus once the rapid conductance change is over, it may take 15 or 20 milliseconds for the membrane potential to return to the base line. When transmitter concentration and conductance change slowly enough for membrane potential to keep up, that is, when the transmitter action is sufficiently longlived for the membrane potential to reflect transmitter concentration accurately at each moment, **transmitter persistence** is said to have occurred.

In the preceding chapters we have contrasted the all-or-none behavior of the action potential with the graded responsiveness seen in PSPs. One might suppose that graded responsiveness implies that PSPs can come in all sizes, ranging continuously from a very small fraction of a millivolt to perhaps tens of millivolts. Although this supposition is for most purposes sufficiently accurate, a more detailed examination of the PSP reveals that its amplitude can increase only in discrete, though very small, steps. It appears that in most, if not all, synapses, transmitter substances are released in packets of nearly constant size (each packet containing many transmitter molecules), and that the total amount released for each PSP must be a multiple of this smallest unit. Thus the smallest possible PSP is one caused by a single packet, the next size being caused by two packets, and so on. The peak amplitude of a PSP resulting from one packet of transmitter is perhaps half a millivolt. Since PSPs add nonlinearly (see page 64), the sizes of the steps caused by increasing the number of packets are not quite equal; the important point is that only those PSP sizes corresponding to the release of an integral number of packets are possible.

If a synaptic terminal can release a variable number of transmitter packets from impulse to impulse, what determines the number of packets released by any given impulse? Since the packets are frequently referred to as **transmitter quanta,** the number of packets causing a PSP is termed its **quantum content.** Our question then is: what determines a PSP's quantum content? The entire answer to this question is not yet known, but certain important factors have been identified. It appears that the release of a packet is a probabilistic phenomenon, that is, in each instant of time there is a certain probability that a packet will be released. Furthermore, it appears that release of the packets is usually independent; the fact that one packet has been released does not influence the release of a neighboring packet. The probability that a packet is released depends on the membrane potential in the axon terminal, and also perhaps on the rapidity of membrane potential changes. If the terminal is depolarized, the probability goes up, and if it is hyperpolarized, the probability goes down. It is also possible that the more rapidly the membrane potential changes, the greater the probability that the transmitter packet will be released. Thus, when an action potential arrives at an axon terminal, the probability that each packet will be released goes from a very low value to a probability of perhaps a half. This means that it is almost certain that a packet will not be released when the terminal is resting, and that it is comparatively probable that a given packet will be released when an impulse arrives at the terminal. It should be noted that the release of transmitter is an all-or-none type of phenomenon, either a packet is released or it is not, and that graded sizes of PSPs result from differing quantum content.

One factor determining the quantum content of a PSP, then, is the probability that a packet of transmitter will be released from the axon terminal. Another factor is the number of packets available to be released. The transmitter packets in a terminal appear to exist in two states, releasable and reserve. The releasable packets are those which can be released by an arriving action potential, and they are continually replenished by the conversion of reserve packets into releasable ones. The number of packets actually released by a nerve impulse, that is, the quantum content of a PSP, is roughly the product of the probability that a single packet will be released and the number available to be released. Thus the quantum content of a PSP can decrease by either decreasing the number available (heavy recent use of the synapse would do this, for instance) or by decreasing the probability of release. Probability of release can be lowered by decreasing the amplitude of the action potential in the axon terminal, and this in turn can be achieved by depolarizing the terminal (see discussion of presynaptic inhibition on page 58). Depolarizing the terminal results in smaller terminal action potentials, probably by causing inactivation of the sodium conductance mechanism and thereby decreasing the total sodium current carrying capacity of the membrane.

In summary, quantum content of a PSP depends on the probability that a transmitter packet will be released and on the number of packets available to be released. Probability increases with depolarization, whereas the number of available packets varies according to the synapse's recent history of use. Available packets are replenished from a reserve, but since this process takes time, the number available to be released often decreases immediately after a PSP and then returns with time to the resting number. The size of the PSP depends on quantum content, postsynaptic membrane potential, and, finally on the state of the postsynaptic membrane; there is some evidence that postsynaptic membranes become insensitive to transmitter if they have been recently flooded with high transmitter concentrations. Although much work remains to be done on the details of synaptic physiology, it can be anticipated that, since this is currently an area of very active research, rapid gains will be made in the coming years.

Chapter 10

QUANTITATIVE THEORIES IN NEUROPHYSIOLOGY

Chapters 1 through 9 have been concerned with a strictly qualitative description of neuronal properties. Because the use of quantitative techniques is so important in certain areas of neurophysiology, this final chapter, making use of these techniques, has been included. Very little of the physiology presented here has not already appeared in one form or another, but the more detailed treatment should give those readers with the appropriate mathematical tools a deeper insight into certain aspects of the physiology dealt with earlier by less appropriate means. Because this chapter is essentially a recapitulation of selected topics, it may be omitted if desired; alternatively, the less mathematically advanced treatments found in Ruch and Patton or Bard (see Suggested References for Chapters 2 through 4) may be substituted. The first two-thirds of this chapter parallels quite closely the qualitative treatment of neuronal mechanisms for the action potential and PSP given in Chapter 9, whereas the final section deals with the origin of potentials in the fluids surrounding active neurons.

THE ACTION POTENTIAL, THE RESTING POTENTIAL, AND PASSIVE SPREAD

We now turn to a systematic (rather than historical) development of the theory of the action potential. This theory takes the form of a set of differential equations describing the time course of the membrane potential, given the appropriate initial conditions; the equations relate membrane potential to certain functions (themselves depending on time and membrane potential) which may be given the physical significance of the axon membrane's permeability to certain ions. Although the eventual goal of this section will be the presentation of the full set of equations for the action potential, it will be necessary first to develop some basic equations upon which later considerations will turn. Our development, in fact, divides conveniently into three parts. The first part deals with general equations relating membrane potential, ionic current flow, and membrane permeabilities. These first equations are derived from the basic **flux equation,** and are conveniently summarized in terms of an electrical circuit, known as the membrane's **equivalent circuit,** which is a model for the electrical properties of the membrane. By taking into account the axon's geometry, the second part of the discussion extends the equivalent circuit description into a general differential equation for membrane potential in a neuron. Membrane potential, in this equation, is seen to depend on unknown functions which may be interpreted as describing the permeability of the membrane to certain ions. At this point it will be clear that understanding the action potential is equivalent to knowing these functions (referred to, for reasons which will become apparent, as **conductance functions,** or simply **conductances**). In the final part of the discussion of the action potential, we shall present the experimental approach used by Hodgkin and Huxley to discover empirically these conductance functions. In the course of this treatment of the action potential, a discussion of the resting potential and passive spread of potential will be included. We turn first, then, to the derivation of the equivalent circuit description of the membrane's electrical properties.

We must first consider the ionic current through a flat square centimeter of membrane. We shall deal with only two ions, sodium and potassium, but a generalization of the equations to a case where more than two ions are important proceeds in a straightforward manner. It is assumed that the behavior of ions within the membrane is governed by the flux equation:

$$J_{\mathrm{Na}} = -u_{\mathrm{Na}}(x)c_{\mathrm{Na}}(x)\left(\mathcal{F}\frac{d\psi(x)}{dx} + \frac{RT}{c_{\mathrm{Na}}}\frac{dc_{\mathrm{Na}}}{dx}\right) \quad \text{(sodium flux)} \qquad (10.1a)$$

$$J_{\mathrm{K}} = -u_{\mathrm{K}}(x)c_{\mathrm{K}}(x)\left(\mathcal{F}\frac{d\psi(x)}{dx} + \frac{RT}{c_{\mathrm{K}}}\frac{dc_{\mathrm{K}}}{dx}\right) \quad \text{(potassium flux)} \qquad (10.1b)$$

The meaning of the symbols is as follows:

J_{Na}	sodium flux through the membrane (moles/second/unit area of membrane)
$u_{\mathrm{Na}}(x)$	the mobility of sodium at position x within the membrane
$c_{\mathrm{Na}}(x)$	the concentration of sodium at x
$\psi(x)$	the electrostatic potential at x
R	the gas constant
T	the temperature (°K)
$\mathcal{F}$	the Faraday
x	distance from the outer surface of the membrane along a normal to the membrane surface (positive direction is into the membrane)

Symbols in the potassium equation have an analogous meaning. The flux equation may be thought of as expressing the assumption that ion flux is proportional to the sum of the driving forces on an ion (concentration and electrostatic potential gradients being the driving forces considered) *times* the number of ions [local concentration $c_{\mathrm{Na}}(x)$] *times* the effect of a unit driving force on the ion [the mobility $u_{\mathrm{Na}}(x)$]. Thermodynamic or statistical mechanical justifications for this equation are possible (although not completely satisfactory); for present purposes, however, we will simply assume it to be correct.

Membrane potential (the inside-outside potential difference), inside and outside ion concentrations, and ionic currents are the variables we are interested in; relations between these variables may be obtained by integrating the flux equations through the membrane. Multiplying both sides of the flux equation (10.1) by

$\mathscr{F}$ (the Faraday) gives the current I_{Na} carried by the sodium ion and the current I_{K} carried by the potassium ion:

$$I_{\text{Na}} = \mathscr{F} J_{\text{Na}} = -u_{\text{Na}} c_{\text{Na}} \mathscr{F}^2 \left(\frac{d\psi}{dx} + \frac{RT}{\mathscr{F} c_{\text{Na}}} \frac{dc_{\text{Na}}}{dx} \right) \qquad \text{sodium current}$$

$$I_{\text{K}} = \mathscr{F} J_{\text{K}} = -u_{\text{K}} c_{\text{K}} \mathscr{F}^2 \left(\frac{d\psi}{dx} + \frac{RT}{\mathscr{F} c_{\text{K}}} \frac{dc_{\text{K}}}{dx} \right) \qquad \text{potassium current}$$

Dividing by $uc\mathscr{F}^2$, integrating from inside of the membrane to its outer surface, and remembering that the outer surface of the membrane is (by convention) at zero potential, we obtain

$$I_{\text{Na}} \int_{\text{in}}^{\text{out}} \frac{dx}{u_{\text{Na}}(x) c_{\text{Na}}(x) \mathscr{F}^2} = E_m - \frac{RT}{\mathscr{F}} \log \frac{c_{\text{Na}}(\text{out})}{c_{\text{Na}}(\text{in})} \qquad (10.2\text{a})$$

$$I_{\text{K}} \int_{\text{in}}^{\text{out}} \frac{dx}{u_{\text{K}}(x) c_{\text{K}}(x) \mathscr{F}^2} = E_m - \frac{RT}{\mathscr{F}} \log \frac{c_{\text{K}}(\text{out})}{c_{\text{K}}(\text{in})} \qquad (10.2\text{b})$$

In this equation, E_m is the electrostatic potential at the inner surface of the membrane, $c_{\text{Na}}(\text{in})$ and $c_{\text{K}}(\text{in})$ are the inside sodium and potassium concentrations, and $c_{\text{Na}}(\text{out})$ and $c_{\text{K}}(\text{out})$ are the outside sodium and potassium concentrations. It is to be especially noted that current has been assumed to be constant through the membrane. That is, the number of sodium (for example) ions entering a small volume equals the number leaving that volume, so that sodium is not collecting or becoming depleted anywhere within the membrane; thus we have assumed that the relaxation times for the internal ionic concentration profiles are so short that the membrane may always be considered to be in a steady state.

By considering the dimensions of the various terms in (10.2) it can be seen that the terms $(RT/\mathscr{F}) \log [c_{\text{Na}}(\text{out})/c_{\text{Na}}(\text{in})]$ and $(RT/\mathscr{F}) \log [c_{\text{K}}(\text{out})/c_{\text{K}}(\text{in})]$ may be considered to be electromotive forces, whereas $\int_{\text{in}}^{\text{out}} \frac{dx}{c_{\text{Na}} u_{\text{Na}} \mathscr{F}^2}$ and $\int_{\text{in}}^{\text{out}} \frac{dx}{c_{\text{K}} u_{\text{K}} \mathscr{F}^2}$ are resistances. We define the following:

$$E_{\text{Na}} = \frac{RT}{\mathscr{F}} \log \frac{c_{\text{Na}}(\text{out})}{c_{\text{Na}}(\text{in})} \qquad g_{\text{Na}} = \frac{1}{\int_{\text{in}}^{\text{out}} \frac{dx}{c_{\text{Na}} u_{\text{Na}} \mathscr{F}^2}}$$

$$E_{\text{K}} = \frac{RT}{\mathscr{F}} \log \frac{c_{\text{K}}\ (\text{out})}{c_{\text{K}}\ (\text{in})} \qquad g_{\text{K}} = \frac{1}{\int_{\text{in}}^{\text{out}} \frac{dx}{c_{\text{K}} u_{\text{K}} \mathscr{F}^2}}$$

Established terminology dictates that E_{Na} be called the sodium equilibrium potential, E_K the potassium equilibrium potential, g_{Na} the sodium conductance, and g_K the potassium conductance. E_m is, of course, the membrane potential. Rewriting (10.2) using these new symbols gives

$$I_{Na} = g_{Na}(E_m - E_{Na}) \quad (10.3a)$$

$$I_K = g_K(E_m - E_K) \quad (10.3b)$$

Equations (10.3) may be represented by an electrical circuit known as the equivalent circuit for the membrane (see Figure 10-1). The equivalent circuit represented in Figure 10-1 is then a formal representation of the nerve membrane in that membrane potential can be determined if total membrane current, membrane conductances, and equilibrium potentials (inside and outside ion concentrations) are known. Equation (10.3) represents restrictions placed on the action potential (in the sense that the action potential must satisfy these equations) by the (assumed) nature of the ionic currents within the membrane.

Axon membranes, together with other plasma membranes, have a lipoprotein structure with a high dielectric constant, and are relatively impermeable to ions of all species. This fact may be included in the equivalent circuit by placing a capacitance in parallel with the other elements as illustrated in Figure 10-2. A number of different cell membranes appear to have capacities on the order of 1 microfarad/cm^2, a very considerable capacitance indeed.

Axons are generally long, thin cylinders. It is thus apparent that the distributed nature of the resistance-capacitance elements must be included in the equivalent circuit. A better approximation to the actual case is presented in Figure 10-3 where a number of equivalent circuits of the type shown in Figure 10-2 are connected with resistors representing the longitudinal resistance of the axoplasm. A still better description is in the form of a differential equation, the **cable equation,** which may be derived as follows. We shall assume for simplicity that the solution bathing the axon is isopotential. Making use of Ohm's law, the longitudinal current i_l in the axon is

$$i_l = \frac{1}{R_{ax}} \frac{\partial E_m(y)}{\partial y}$$

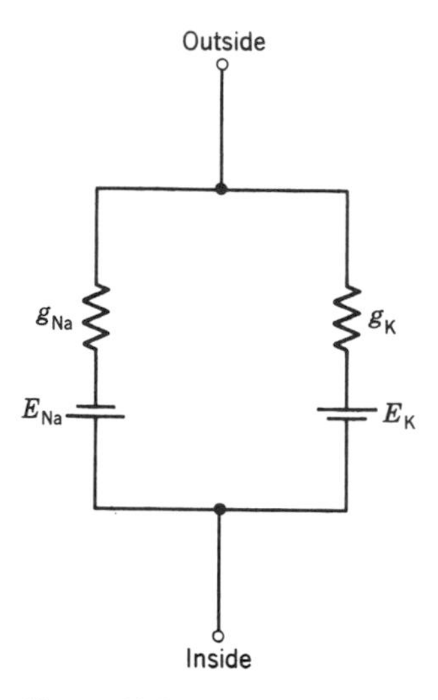

Figure 10-1. Equivalent circuit for a square centimeter of membrane permeable only to potassium and sodium. The conductances (g_{Na} and g_K) and the electromotive forces (E_{Na} and E_K) are defined in the text.

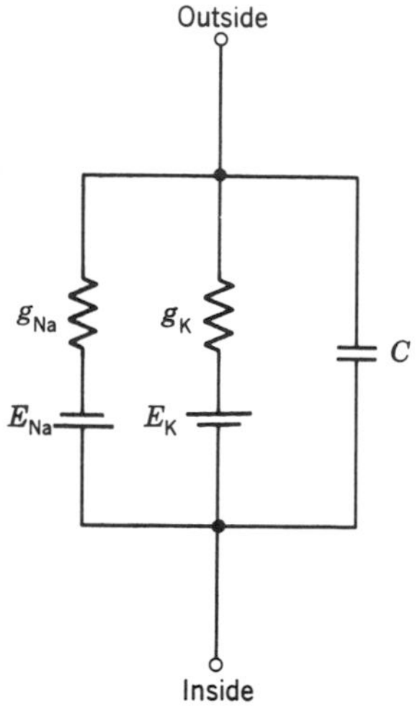

Figure 10-2. Equivalent circuit for membrane incorporating the membrane capacitance.

where R_{ax} is the axoplasmic resistance per unit length of axon, E_m is the membrane potential, and y is the distance along the axon. Since all current must flow either longitudinally or through the membrane, the membrane current i_m is given by the change in longitudinal current with distance

$$i_m = \frac{\partial i_l}{\partial y} = \frac{1}{R_{ax}} \frac{\partial^2 E_m}{\partial y^2} \tag{10.4}$$

The preceding is an expression for membrane current; this membrane current in turn has two components, capacitive current and ionic current (here we assume that the most important ion current carriers are sodium and potassium)

$$i_m = \frac{1}{R_{ax}} \frac{\partial^2 E_m}{\partial y^2} = C \frac{\partial E_m}{\partial t} + I_{Na} + I_K \tag{10.5}$$

In this equation C represents the membrane capacity per unit length of axon, and I_{Na} and I_K are the sodium and potassium currents per unit length. Substitution from equations (10.3) for the sodium and potassium current gives finally

$$\frac{1}{R_{\mathrm{ax}}}\frac{\partial^2 E_m}{\partial y^2} = C\frac{\partial E_m}{\partial t} + g_{\mathrm{Na}}(E_m - E_{\mathrm{Na}}) + g_{\mathrm{K}}(E_m - E_{\mathrm{K}}) \quad (10.6)$$

Any change in membrane potential must satisfy this equation, and in particular, the action potential must satisfy the equation. Since E_{Na} and E_{K} are fixed by the inside and outside ion concentrations (these do not change appreciably with time), membrane potential variations as seen in the action potential can arise only through variations in the ionic conductances or in the membrane capacitance. From experiment we know that conductance changes accompany the action potential, but that the membrane capacity remains constant. Thus, we need only know what functions to substitute for g_{Na} and g_{K} into equation (10.6) in order to have a description of the action potential. We may say, then, that the key to understanding the action potential is an understanding of the ionic conductance (or equivalently, permeability) changes, for these underlie the membrane potential variations observed in experiments.

Before describing the experimental procedure for the determination of the sodium and potassium conductance functions (g_{Na} and g_{K}), it is well to pause briefly to explore several implications of equation (10.6). From this equation, together with a few other facts, it is possible to obtain some insight into the mechanism of the resting potential and the phenomenon of passive spread. From experiments such as those described in Chapter 2, we know that in the resting axon, the membrane potential is constant over

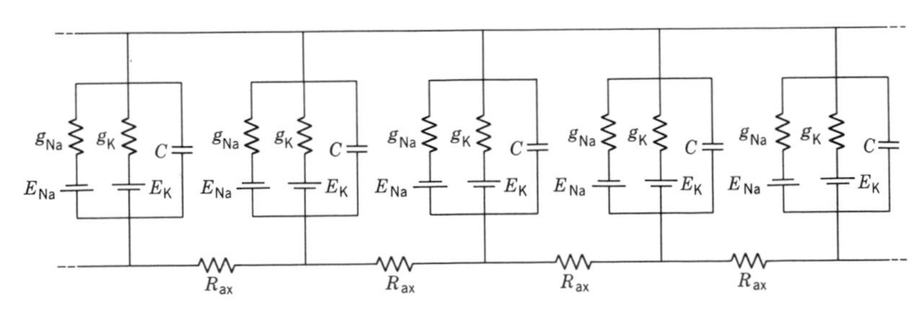

Figure 10-3. Model for a length of axon which takes into account the longitudinal resistance of the axon.

both time and distance: there are no important temporal fluctuations in the resting potential, and it has essentially the same value everywhere within the neuron. Thus, for the resting neuron, $\partial^2 E_m/\partial y^2$ and $\partial E_m/\partial t$ vanish, and equation (10.6) becomes simply

$$g_{\mathrm{Na}}^{*}(E_m^{*} - E_{\mathrm{Na}}) = -g_{\mathrm{K}}^{*}(E_m^{*} - E_{\mathrm{K}})$$

Here, we have starred E_m, g_{Na} and g_{K} to indicate that they represent the resting values of these variables. Solving this equation for E_m^{*},

$$E_m^{*} = \frac{g_{\mathrm{Na}}^{*}}{g_{\mathrm{Na}}^{*} + g_{\mathrm{K}}^{*}} E_{\mathrm{Na}} + \frac{g_{\mathrm{K}}^{*}}{g_{\mathrm{Na}}^{*} + g_{\mathrm{K}}^{*}} E_{\mathrm{K}} \qquad (10.7)$$

we find that the resting potential depends on the sodium and potassium equilibrium potentials, and the sodium and potassium conductances. In general, of course, one would have to include chloride, calcium, and conductances and equilibrium potentials for any other ions present; we have assumed for notational simplicity (and in anticipation of results to be discussed later) that the only ions whose conductances are appreciable are sodium and potassium. Note from equation (10.7) that any ion whose conductance is a negligible fraction of the total conductance contributes negligibly to the resting potential. Note further that since $E_{\mathrm{Na}} = (RT/\mathscr{F}) \log [C_{\mathrm{Na}}(\mathrm{out})/C_{\mathrm{Na}}(\mathrm{in})]$ and the inside sodium concentration is less than the outside, high resting sodium conductance would make the resting membrane potential positive with respect to the outside, whereas (by the same argument) a high resting potassium conductance would make the resting potential negative. The resting potential is in fact negative, and so we may surmise that (to the extent that sodium and potassium have the only appreciable fluxes) the resting potential is mainly determined by a relatively larger potassium permeability and smaller sodium permeability. From a type of experiment to be described later (voltage-clamp experiment) we know that this explanation for the resting potential is indeed accurate.

Since a sufficiently large depolarization results in the production of an action potential, it is to be anticipated that the conductance functions depend on membrane potential. For appropriately small membrane potential deviations, however, it is also to be anticipated that the conductances will be (approximately) constant. Thus the passive circuit properties of the neuron discussed in Chapter 2 should be described by equation (10.6) with the conductances taken as constants (independent of membrane potential and time). Since this description is valid only for a restricted range of membrane potential around the resting potential, it is convenient to change variables in (10.6) so that membrane potentials are considered as deviations from the resting potential. Defining $V = E_m - E_m^*$, it can be verified (remembering that E_m^* is a constant, and making use of equation 10.7) that equation (10.6) becomes

$$\frac{1}{R_{ax}} \frac{\partial^2 V(y, t)}{\partial y^2} = C \frac{\partial V(y, t)}{\partial t} + (g_{Na}^* + g_K^*) V(y, t) \qquad (10.8)$$

Here we have again placed stars on the sodium and potassium conductances to emphasize that they are constant (at their resting values). This equation is the well-known cable equation describing the behavior of (for example) an undersea telegraph cable with a resistive core and a leaky insulator. Equation (10.8) is the quantitative expression of the passive properties of the axon described in Chapter 2 (passive spread of potential, and blunted voltage response to square current steps).

It is possible, but usually somewhat tedious, to solve equation (10.8) for a variety of boundary and initial conditions representing, for example, the voltage response of a neuron to current applied through an intracellular electrode. Rather than considering solutions of the complete equation here, however, we shall indicate the behavior of the equation by considering two special cases; the first treats a neuron with an isopotential interior, while the second deals with the steady-state potential distribution for a point depolarization of a cylindrical axon or dendrite. If a neuron has an (approximately) isopotential interior, the membrane current must equal the current applied (through an intracellular electrode, for example), and the term on the left-hand side of (10.8) is replaced by the applied current

$$i_{app} = C \frac{dV(t)}{dt} + (g_{Na}^* + g_K^*) V(t) \qquad \text{Response of a neuron with an isopotential interior} \qquad (10.9)$$

Equation (10.9) will be recognized as the equation of a parallel

RC circuit with a **time constant** $(C/g^*_{Na} + g^*_K)$; the response of such a circuit to a square current pulse is illustrated in Figure 2-4 of Chapter 2. In order to obtain (10.9) we have neglected the distributed nature of the membrane resistance and capacitance, and approximated the axon electrical properties by a lumped circuit. Although not very accurate for a real axon or neuron, the lumped circuit model embodied in equation (10.9) is qualitatively correct, and for many purposes, of sufficient quantitative precision as a model for the neuron's passive electrical properties. For example, the response of the neuron's soma to applied current can sometimes be successfully approximated by equation (10.9), although in general, of course, the equation is quite inaccurate.

If a step current is applied to an axon through an intracellular electrode, the membrane potential will pass through a transient phase to finally reach a steady-state (assuming that a constant current is passed for a sufficiently long time, and that the shifts in membrane potential are well below threshold); as a rough approximation, the transient will be described by equation (10.9). Once the steady-state is reached, the membrane potential is constant over time, and so the partial of voltage with respect to time in equation (10.8) vanishes to give

$$\frac{1}{R_{ax}} \frac{d^2V(y)}{dy^2} = (g^*_{Na} + g^*_K)V(y) \tag{10.10}$$

Here, then, is an ordinary second-order differential equation for membrane potential as a function of distance along the axon. The general solution to (10.10) is given by

$$V(y) = A \exp\left[\sqrt{R_{ax}(g^*_{Na} + g^*_K)}\; y\right] + B \exp\left[-\sqrt{R_{ax}(g^*_{Na} + g^*_K)}\; y\right]$$

where A and B are constants to be determined from the boundary conditions and $1/\sqrt{R_{ax}(g^*_{Na} + g^*_K)}$ is known as the **space constant.** We shall denote the membrane potential at the site of the current-passing electrode by $V(0)$, choosing this point as the origin of our coordinate system. Further supposing that the axon extends to infinity on either side of the current-passing electrode, we have the boundary conditions

$$V(+\infty) = V(-\infty) = 0$$

Considering first values of y for which $y \geqslant 0$, the boundary conditions are met by letting $A = 0$ and $B = V(0)$ to give the solution:

$$V(y) = V(0) \exp\left[-\sqrt{R_{ax}(g_{Na}^* + g_K^*)}\, y\right] \qquad y \geqslant 0$$

For negative y it is necessary to choose $A = V(0)$ and $B = 0$ in order that $V(-\infty)$ vanish:

$$V(y) = V(0) \exp\left[+\sqrt{R_{ax}(g_{Na}^* + g_K^*)}\, y\right] \qquad y \leqslant 0$$

These solutions may be combined to give the steady-state depolarization as a function of distance from a point source of current (located at $y = 0$):

$$V(y) = V(0) \exp\left[-\sqrt{R_{ax}(g_{Na}^* + g_K^*)}\, |y|\right] \qquad (10.11)$$

The graph of equation (10.11) was presented in Figure 2-10*b* of Chapter 2 as part of the qualitative description of passive spread. Although this equation applies to the steady state, it nevertheless gives a good qualitative picture of the dependence of membrane potential on distance from a current source even for non-steady-state situations.

Before proceeding further, it is necessary to emphasize the fact that any appearance of rigor in the preceding development is more illusory than real. A number of approximations have been made which are of no essential importance. For example, the external resistance of the solution bathing the axon has been neglected, and concentrations rather than activities appear in the flux equation (10.1). These and similar approximations do, of course, change the numerical predictions made by the various equations, but they do not alter the form of the equations. The assumption of a steady-state (p. 138) is also probably not in serious error (see the Cohen and Cooley paper cited in the references for this chapter). There is, however, an assumption central to the entire preceding treatment which is not so inconsequential. Whereas the flux equation is indeed valid for many situations, it is not known whether or not it is applicable to ions within the neuron membrane. Strictly speaking, in fact, the flux equation is not even meaningful in a system such as the plasma membrane. The flux equation deals with macroscopic (average) variables such as concentration, whereas the nerve membrane is so thin that a more detailed microscopic treatment is required. For example, it is frequently assumed that ions pass virtually single file through pores in the membrane, in which case a pore would contain on the order of ten ions. Clearly, to integrate concentration through the membrane in such a case, as was done on page 138, is suspect. This is not to say that final results obtained through the flux equation are necessarily invalid, but only that they are not inevitably correct. The main point is that the preceding development of the membrane equivalent circuit starting with the flux equation is best viewed as a plausability argument rather than a rigorous derivation.

The argument to this point may be summarized as follows: beginning with the flux equation governing the behavior of ions within the membrane, an equivalent circuit for the membrane was derived. This circuit formally relates ionic conductances, inside-outside ion concentrations, membrane potential, and ionic current through the membrane to one another. The equivalent circuit was completed by adding a parallel capacitance; justification for the capacitance rests finally on empirical observation, but is to be anticipated from the fact that nerve membrane is a thin, sparingly permeable, high-dielectric constant material separating two good conductors. Because of the neuron's geometry, further restrictions are placed on the relations between ionic conductances, current, and membrane potential. Combining the equivalent circuit with geometrically imposed restrictions, we finally arrived at the general equation governing the electrical behavior of the neuron (assuming appropriate boundary and initial conditions are supplied). From this equation (together with the experimental fact that membrane capacitance is constant) it can be seen that ionic conductance changes must underlie the action potential. When we consider special cases of this general equation, explanations result for the resting potential in terms of resting conductances and for passive spread in terms of solutions to the general equation for situations in which conductances are constant. We turn now to the method used by Hodgkin and Huxley to determine the conductance functions, and thus to achieve a quantitative description of the action potential.

The general equation for the action potential is

$$\frac{1}{R_{ax}} \frac{\partial^2 E_m(y, t)}{\partial y^2} = C \frac{\partial E_m(y, t)}{\partial t} + g_{Na}(E_m(y, t) - E_{Na}) + g_K(E_m(y, t) - E_K) \qquad (10.6)$$

where g_{Na} and g_K represent the unknown conductance functions; it is these functions which must be specified to complete the description of the action potential. Since the initiation of the nerve impulse depends on displacing the neuron's membrane potential to a critical level, and since we have concluded that conductance changes underlie the action potential, we may anticipate that the conductances will be functions (at least) of membrane potential and time

$$g_{Na} = g_{Na}(E_m, t) \qquad g_K = g_K(E_m, t)$$

To determine these functions we must have relationships between them and other variables which can be controlled or measured. Equation (10.6) is such a relationship between conductance and membrane potential, but unfortunately it is too complicated: a second partial of membrane potential with respect to distance and a first partial of membrane potential with respect to time appear, and there are (in our simplified case) two conductance functions which must be separated. Clearly some experimental method of simplifying equations is desirable. It is, in fact, experimentally possible to remove the time and space derivatives appearing in equation (10.6), and, in addition, separate the conductances.

The second partial of membrane potential with respect to distance (along the length of the axon) arises from longitudinal current flows, and can thus be eliminated if the interior of the axon can be made isopotential. Experimentally this can be easily accomplished (in the squid giant axon) by inserting a wire longitudinally within the axon to effectively reduce longitudinal resistance to zero (Figure 10-4). For this case, membrane current is exactly whatever current is applied from external sources:

$$i_{app} = i_m = C \frac{\partial E_m}{\partial t} + g_{Na}(E_m - E_{Na}) + g_K(E_m - E_K) \qquad (10.12)$$

(As indicated in the figure, a circular external ground electrode short-circuits any longitudinal resistance of the bathing solution

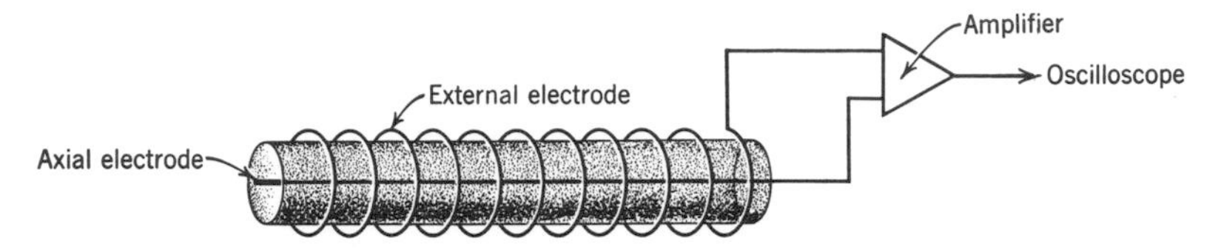

Figure 10-4. Arrangement for space clamping a length of axon.

and insures that all current flow is normal to the membrane.) The procedure just described for removing the space derivative from the equation is known for reasons which will later become apparent as the **space-clamp.**

A procedure, known as **voltage clamping,** is also available for removing the time derivative in equation (10.12); this is accomplished by constraining the membrane potential from varying through the use of special electronic equipment. If a current is passed through the space-clamped axon membrane, the membrane voltage will change to a new value. However, because the conductances depend on membrane potential, in general, they will change, causing membrane potential to vary in accordance with equation (10.12). The principle of the voltage-clamp is to measure membrane potential and supply the axon membrane (through an internal electrode) with whatever current is necessary to maintain membrane potential at the desired (constant) value. Voltage clamping is (in principle) accomplished by a circuit such as that illustrated in Figure 10-5. A high gain differential amplifier supplies sufficient current to maintain (essentially) no difference between the potential of the interior of the axon and the second input of the amplifier to which the desired voltage for the membrane potential, known as the *clamping voltage*, is applied. By applying only step changes to the second input of the amplifier, the membrane potential of the axon is always constant except for the rapid transitions between constant voltage levels. Except for these brief capacitative transients, then, all current is ionic:

$$i_{app} = i_m = g_{Na}(E_m - E_{Na}) + g_K(E_m - E_K) \qquad (10.13)$$

Current-voltage relation for voltage clamped axon

This ionic current is, of course, easily measured as indicated in Figure 10-5. Except for the question of separating the conductances in equation (10.13), the problem of determining the conductance functions is, in principle, straightforward: E_m is under the control of the experimenter, i_m is measurable, and the unknown conductance functions relate E_m to i_m in the simple way exhibited in equation (10.13).

Before considering the method used for separating the conductances, we shall describe the type of result obtained from voltage-clamp experiments. When the membrane potential is caused to change from the resting level to (say) 40 millivolts depolarized, there results a brief capacitance current, a relatively shortlived inward current, and finally a prolonged outward current lasting for as long as the axon is maintained in the depolarized state (Figure 10-6, *40*). For other magnitudes of depolarizations, qualitatively the same sequence of events occurs, except that with a sufficiently large depolarization, the early outward current becomes inward (Figure 10-6). Assuming, as we have been for convenience, sodium and potassium ions are those carrying virtually all of the current, these results can be readily interpreted by reference to equation (10.13). This interpretation will be facilitated by putting in numbers for the sodium and potassium batteries in the equivalent circuit. From sodium and potassium inside-outside concentrations, we find that $E_{\mathrm{Na}} \approx +55$ millivolts

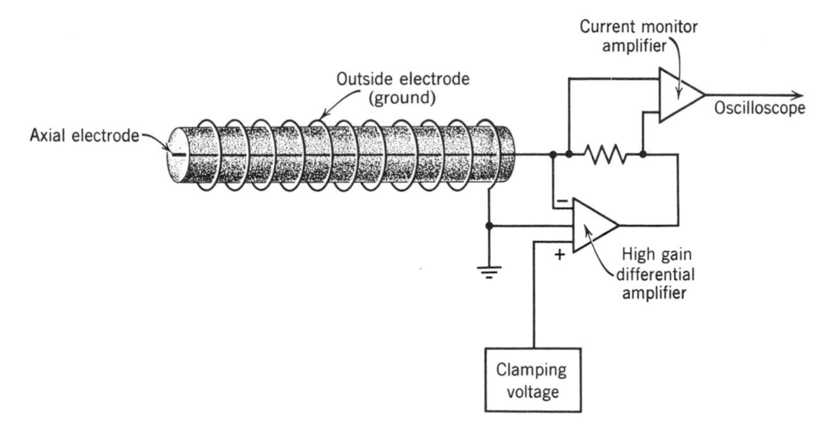

Figure 10-5. Arrangement for voltage clamping a length of axon.

and $E_K \approx -72$ millivolts. Thus the equation becomes

$$i_{app} = i_m = g_{Na}(E_m - 55) + g_K(E_m + 72) \qquad (10.14)$$

For depolarizations up to $E_m = +55$ millivolts, sodium current would be inward, whereas potassium current would be outward; the net current, of course, depends on the relative magnitude of the sodium and potassium conductances. Evidently, the initial inward current in the voltage-clamp is a large sodium current

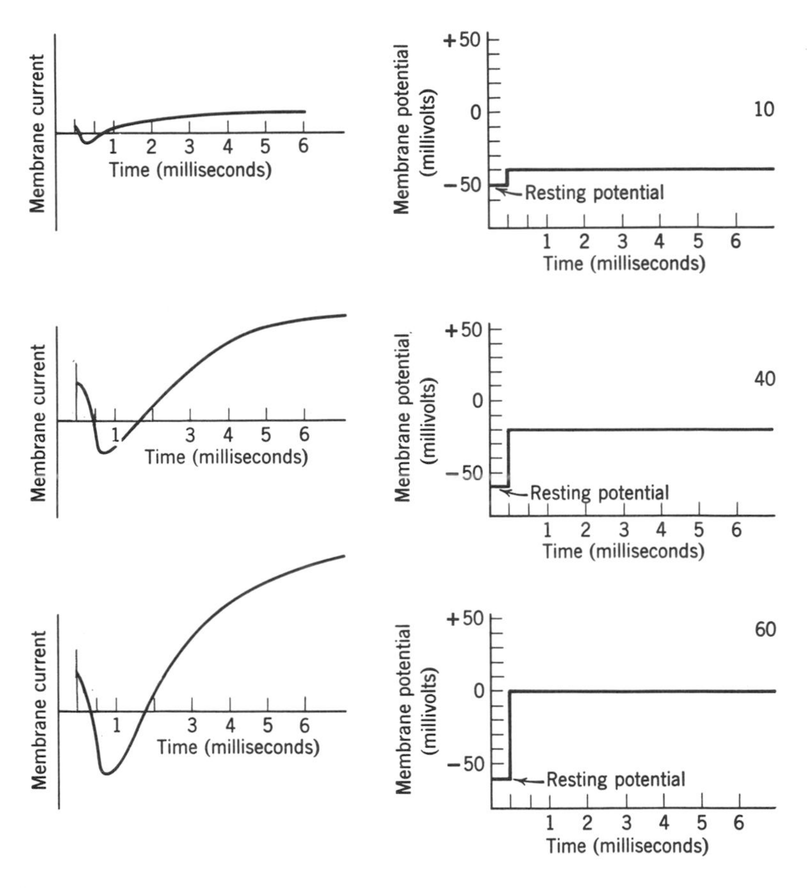

Figure 10-6. Records from an hypothetical voltage clamping experiment. The membrane potential (as a function of time) is given in the right-hand graphs, and the axon membrane current in the left-hand graphs. Number next to each voltage graph indicates the magnitude of the depolarization used in clamping.

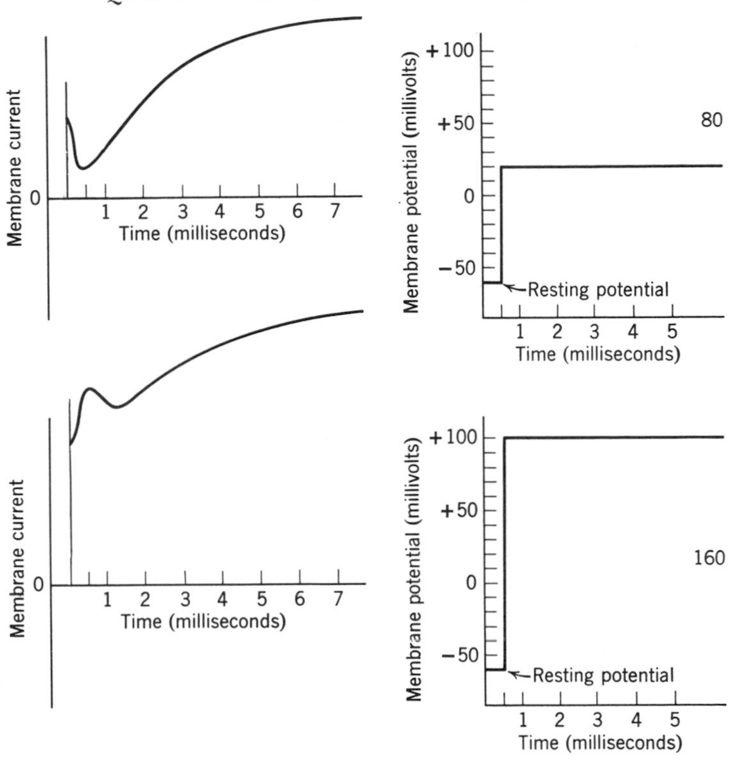

Figure 10-6 (*continued*).

reflecting a dominant sodium conductance, whereas the later outward current is due to potassium flux and a large potassium conductance. For depolarization exceeding $E_m = +55$ millivolts, equation (10.14) demands that both sodium and potassium currents be outward; hence the two-stage outward current pattern for these larger depolarizations, the initial stage reflecting the increased sodium conductance and the later stage resulting from the large potassium conductance. The question now remains how the two (or, in general, more) conductances can be separated.

Sodium and potassium conductances have been found to be independent of one another, a fact which greatly simplifies the task of separating the conductances in equation (10.13). By replacing the external sodium with choline, a cation which carries no appreciable current through the membrane, equation (10.13) further simplifies to

$$i_m = i_K = g_K(E_m - E_K)$$

or

$$g_K = \frac{i_K}{E_m - E_K}$$

Voltage-clamp records with choline replacing sodium show, indeed, the absence of the initial inward current (Figure 10-7). From current records for a series of different clamping voltages, we obtain through the preceding equation a family of potassium conductance functions of time with membrane potential appearing as a parameter (Figure 10-8*a*).

If the same clamping voltages are used with sodium present and then with sodium replaced by choline, the sodium current can be found by subtraction:

$$i_m - i_K = i_{Na} = g_{Na}(E_m - E_{Na})$$

or

$$g_{Na} = \frac{i_{Na}}{E_m - E_{Na}}$$

Again, we find a family of sodium conductances as a function of time with membrane potential appearing as a parameter (Figure

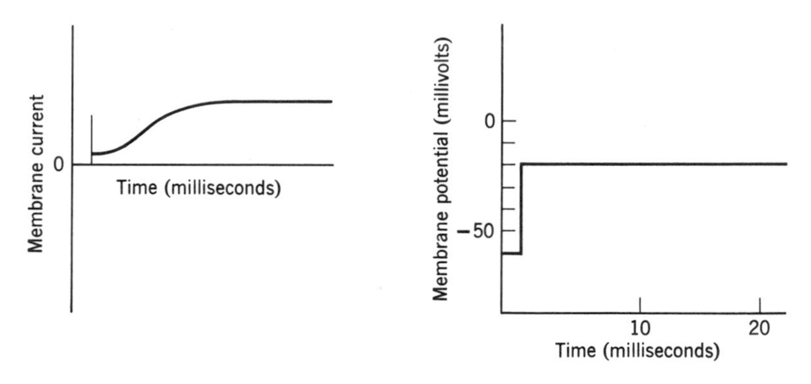

Figure 10-7. Record from hypothetical voltage clamping experiment in which sodium in the bathing solution was replaced by choline. A 40 millivolt step in membrane potential was used.

10-8*b*). It should be noted that whereas potassium conductance increases slowly to a maintained steady-state value during a depolarization, sodium conductance increases more rapidly and then decreases. The decrease in sodium conductance with maintained depolarization is termed **sodium inactivation,** and the conductance increases are called **potassium** and **sodium activation.**

Having determined the empirical conductance functions, Hodgkin and Huxley then found empirical equations describing these functions. The pair of equations describing the potassium conductance are

$$g_K = \bar{g}_K n^4 \tag{10.15}$$

$$\frac{dn}{dt} = \alpha_n(1 - n) - \beta_n n \tag{10.16}$$

where $\bar{g}_K$ is the maximum potassium conductance (a constant), α_n and β_n are rate constants depending on voltage but not on time, and n is the variable characterizing the state of potassium activation. Equations (10.15) and (10.16) were selected arbitrarily to give a reasonably satisfactory fit to the conductance functions determined empirically in the manner described previously. The dependence of the rate constants on voltage is found by plotting the rate constants necessary for each conductance curve (specified by the voltage parameter) as a function of the clamping voltage; empirical curves which satisfactorily fitted the data were then selected. Hodgkin and Huxley found the following functions adequate for their data, although individual axons have somewhat different values of the constants

$$\alpha_n = \frac{0.01(V + 10)}{e^{(V+10)/10} - 1} \tag{10.17}$$

$$\beta_n = 0.125e^{V/80} \tag{10.18}$$

(Here, as before, V denotes the displacement of the membrane potential from its resting value.) Equations (10.15) through (10.17) then are descriptions of the conductance functions seen under voltage clamp conditions.

It will be recalled that the sodium conductance increased when the axon was depolarized, and then decreased again in the pres-

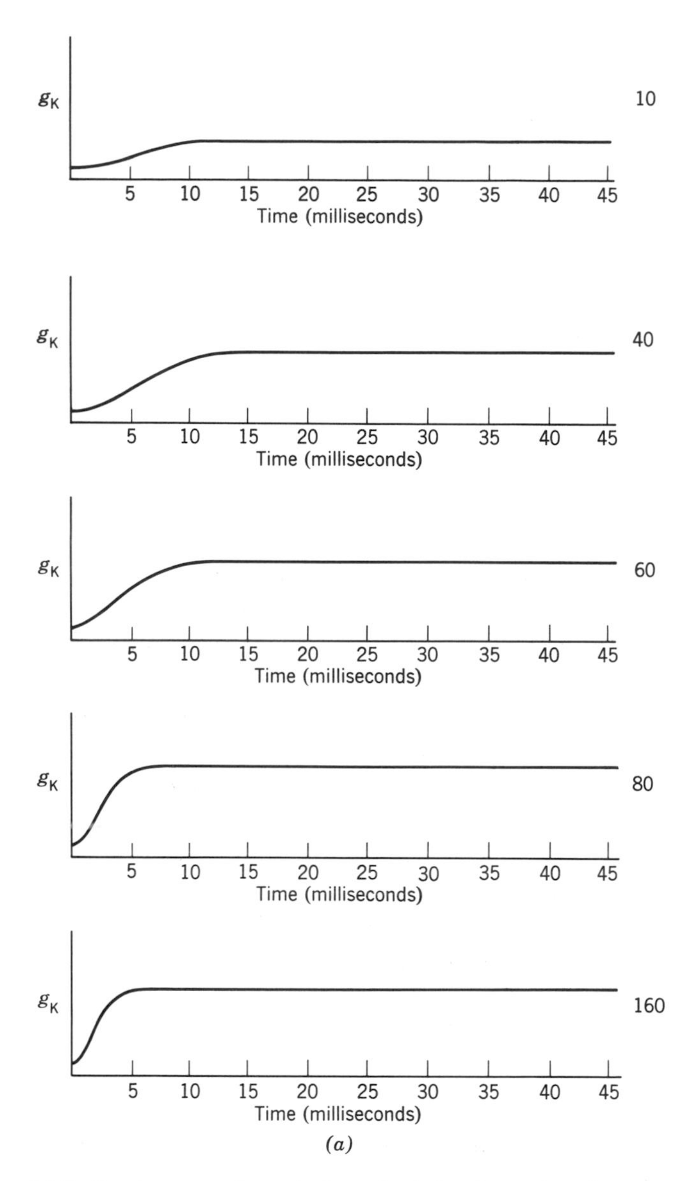

Figure 10-8. Conductances from an hypothetical voltage clamping experiment. Numbers beside the graphs indicate the clamping voltage in millivolts (all depolarizations). (*a*) Potassium conductance. (*b*) Sodium conductance. Note that the time scale on the potassium conductance curves is more compressed than that on the sodium conductance curves.

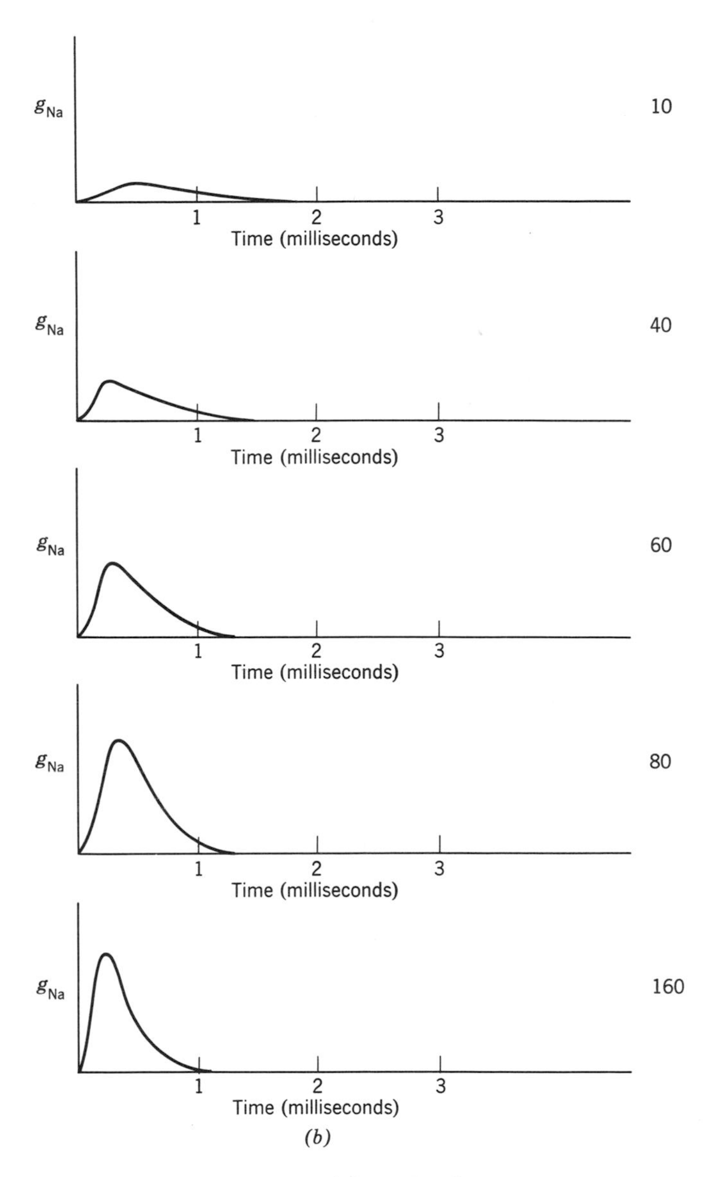

(*b*)

Figure 10-8 (*continued*).

ence of steady depolarization, the phenomenon referred to as sodium inactivation. Because of inactivation, the empirical equations describing sodium conductance contain two variables, m specifying the state of sodium activation, and h, specifying the state of sodium inactivation. As with the potassium activation variable n, it is required that m and h vary in the interval 0 to 1. The empirical equations found adequate to describe the sodium conductance changes found in voltage-clamp experiments are

$$g_{\text{Na}} = \bar{g}_{\text{Na}} m^3 h \tag{10.19}$$

$$\frac{dm}{dt} = \alpha_m (1 - m) - \beta_m m \tag{10.20}$$

$$\frac{dh}{dt} = \alpha_h (1 - h) - \beta_h h \tag{10.21}$$

As before, α_m, β_m, α_h, and β_h are rate constants depending on voltage (but not on time), and $\bar{g}_{\text{Na}}$ is the maximum permissible value of the sodium conductance. The dependance of α_m and β_m on voltage is found to have the same form as for potassium:

$$\alpha_m = \frac{0.1(V + 25)}{e^{(V+25)/10} - 1} \tag{10.22}$$

$$\beta_m = 4e^{V/18} \tag{10.23}$$

A somewhat different form is required for α_h and β_h:

$$\alpha_h = 0.07e^{V/20} \tag{10.24}$$

$$\beta_h = \frac{1}{e^{(V+30)/10} + 1} \tag{10.25}$$

Equations (10.19) through (10.25) represent an adequate empirical description of the sodium conductance as determined by the voltage clamp experiments.

The conductance functions may now be substituted into the general differential equation for the action potential to complete the description; we collect all of the equations for describing the action potential here

$$\frac{1}{R_{\text{ax}}} \frac{\partial^2 E_m}{\partial y^2} = C \frac{\partial E_m}{\partial t} + \bar{g}_{\text{Na}} m^3 h (E_m - E_{\text{Na}}) + \bar{g}_{\text{K}} n^4 (E_m - E_{\text{K}}) \tag{10.25}$$

$$\frac{dn}{dt} = \alpha_n(1 - n) - \beta_n n \tag{10.16}$$

$$\frac{dm}{dt} = \alpha_m(1 - m) - \beta_m m \tag{10.20}$$

$$\frac{dh}{dt} = \alpha_h(1 - h) - \beta_h h \tag{10.21}$$

$$\alpha_n = \frac{0.01(V + 10)}{e^{(V+10)/10} - 1} \tag{10.17}$$

$$\alpha_m = \frac{0.1(V + 25)}{e^{(V+25)/10} - 1} \tag{10.22}$$

$$\alpha_h = 0.07e^{V/20} \tag{10.24}$$

$$\beta_n = 0.125e^{V/80} \tag{10.18}$$

$$\beta_m = 4e^{V/18} \tag{10.23}$$

$$\beta_h = \frac{1}{e^{(V+30)/10} + 1} \tag{10.25}$$

($V = E_m - E_m^*$, where E_m^* is the resting potential.) These equations together constitute a quantitative description of the action potential. It is necessary to check that the conductance functions obtained by the voltage clamp procedure are indeed the correct ones. This is accomplished by obtaining constants (the ones above, for example) from voltage-clamp experiments, and then solving (by approximation methods) the equations for the action potential. If the conductance functions determined by the voltage-clamp procedure are correct, it is to be expected that the form of the action potential should be accurately predicted from (10.25) and the other equations. Satisfactory agreement between predicted and experimentally observed action potentials is indeed found. Furthermore, many properties, such as refractory period, the strength-latency relationship, and the threshold, among others, are adequately predicted by the theory. Altogether, good quantitative agreement between predictions of the theory and experimental observations is consistently obtained, although constants and even the form of some of the accessory equations (for the rate of constants, for example) must be modified for various types of axons, or even different axons from the same species.

The theory of the action potential today is not essentially different from that which is described in the preceding paragraphs, and our treatment is quite similar to the one presented by Hodgkin and Huxley in 1952. The central problem remains the physical basis for the conductance changes, or more generally, the treatment of ion movements within the membrane.

POSTSYNAPTIC POTENTIALS

Having considered the theory of nerve impulse generation, we turn to the problem of dendritic properties. In this section, as in the preceding one, we shall first consider the ionic mechanism underlying the membrane potential variations seen in the PSP.

The derivation of equation (10.6) was perfectly general (if heuristic), and so must apply equally well to the PSP. That is, the PSP must be described by

$$\frac{1}{R_D}\frac{\partial^2 E_m}{\partial y^2} = C\frac{\partial E_m}{\partial t} + g_{\mathrm{Na}}(E_m - E_{\mathrm{Na}}) + g_{\mathrm{K}}(E_m - E_{\mathrm{K}}) + g_{\mathrm{Cl}}(E_m - E_{\mathrm{Cl}}) \tag{10.26}$$

where we have now included (unknown) conductance functions for chloride as well as for potassium and sodium and we have denoted the dendrite's longitudinal resistance by R_D. Although the form of the basic equation is the same for both the PSP and the action potential, there are, of course, a number of important differences in application in the two cases. First, where the conductances are functions of membrane potential for the axon, they probably do not depend on membrane potential for the synaptic membrane, but rather on transmitter concentration in the vicinity of the postsynaptic membrane. Second, the membrane properties (as reflected by the conductance functions) are not spatially homogenous for the dendrite, so that the conductances are functions of distance along the dendrite as well as of transmitter concentration; this is a statement of the fact that the synapse is a relatively small, circumscribed structure.

To complete the description of PSP, we would have to know how the various conductances depend on transmitter concentration (and, perhaps, on time) and also understand the laws governing release, destruction, and diffusion of transmitter. Largely because of the difficulties encountered in attempting to perform the relevant experiments, the information available about conductance functions, and transmitter release and destruction is not very extensive. Transmitter release from axon terminals appears to occur in multiples of a common unit, and the amount released per impulse probably depends on average axon terminal membrane potential and/or action potential amplitude (see page 132). The duration of transmitter action varies greatly from cell to cell. In some neurons, transmitter destruction is rapid, and the total duration of action is on the order of a millisecond, whereas in other neurons transmitter acts for hundreds of milliseconds (the phenomenon known as **transmitter persistence**). Unfortunately, the relation between transmitter concentration and magnitude of conductance change is not known precisely as yet, although it is known to be an increasing monotonic function. Which conductances will change depends, of course, on the type of synapse involved. At excitatory synapses, the available evidence indicates that both sodium and potassium conductances increase with perhaps no change in chloride conductance. At mammalian inhibitory synapses, sodium conductance appears to remain unchanged, but potassium and chloride conductances increase.

For several idealized but nevertheless instructive situations, it is possible to use equation (10.26) to predict the form of the PSP. If the duration of transmitter action is short compared to the membrane time constant, it is reasonable to consider the conductance functions (sodium and potassium in the case of the EPSP, for example) as Dirac delta functions in time and space

$$g_K(y, t) = g_K^* + a_K\, \delta(y)\, \delta(t)$$

$$g_{Na}(y, t) = g_{Na}^* + a_{Na}\, \delta(y)\, \delta(t)$$

Here $g(y, t)$ is conductance as a function of time and position on a cylindrical dendrite, $y = 0$ is the location of the synapse, $t = 0$ is the time of transmitter application, g^* is the resting conductance, and a is a constant representing the magnitude of the conductance change. For this case, the form of the PSP is determined completely by the cable properties of the neuron. Figure 10-9 illustrates the behavior of equation (10.26) for the above conductance functions and an infinite cylindrical dendrite.

In the preceding example we considered one limiting case, namely a duration of transmitter action which is short compared to the membrane time constant. Very slow, rather than very rapid, transmitter destruction provides the other limiting situation. If the duration of transmitter action is long compared to the mem-

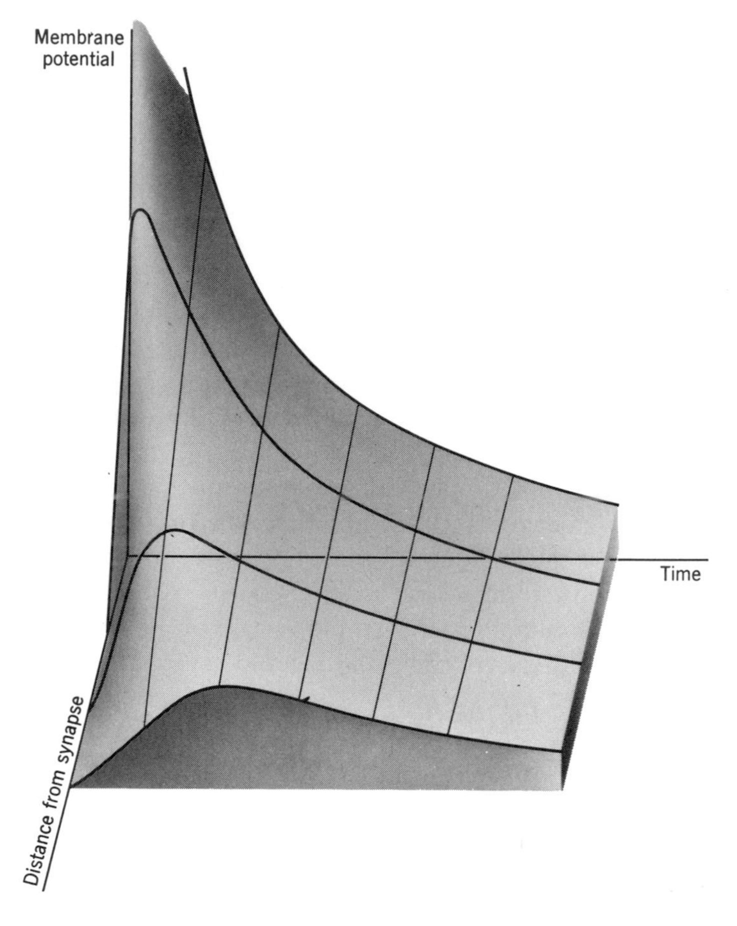

Figure 10-9. PSP produced in a long dendrite by a very brief conductance change. The axis projecting out of the surface of the page represents distance along the dendrite from the site of the conductance change.

brane time constant, membrane potential changes will, except perhaps for the rising phase of the PSP, be sufficiently slow that the $\partial E_m/\partial t$ term in equation (10.26) may be neglected; in other words, the steady-state situation discussed on page 144 obtains at all times, and the falling phase of the PSP has the same shape at all distances from the synaptic region although it is, of course, exponentially attenuated with distance from the synapse according to (10.11). To further evaluate the long transmitter persistence situation, we must have more information about transmitter concentration as a function of time after the nerve impulse arrived at the synapse, and we must also know the relationship between transmitter concentration and the conductances. The assumption that transmitter destruction is a first-order reaction does not do violence to existing information, and it is thus anticipated that concentration will decline approximately exponentially from a rapidly attained peak value (assuming that destruction is not so slow that an appreciable amount of transmitter is not lost from the synaptic region by diffusion). Furthermore, over a 4 or 5 millivolt range, membrane potential displacement is proportional to transmitter concentration. Thus with long transmitter persistence, we anticipate that the form of the PSP would be a fairly rapid rise and a long exponential decay to the resting potential. As was pointed out earlier, one in fact finds a large range of PSP durations in the nervous system. The above predictions are in semiquantitative agreement with the observed PSP forms, although the extent of the agreement is at present unknown since detailed and systematic investigations of PSP form have not yet been done.

INTERPRETATION OF EXTRACELLULAR POTENTIALS

The importance of information obtained by recording neuron activity extracellularly was indicated in Chapter 7; interpretation of these extracellular potential changes depends on an understanding of potentials in a volume conductor. Although the exact treatment of the electromagnetic fields surrounding a neuron is almost hopelessly complex, the approximate theory presented in the remainder of the chapter is less difficult, and leads also to a semiquantitative understanding of the extracellular signs of neuronal activity.

Before entering into the mathematical theory of potentials in a volume conductor, we shall first summarize briefly some of the results. When a PSP or an action potential occurs in a neuron, ionic currents flow through the neuron membrane and the surrounding tissue; associated with these currents are potential gradients in the region around the neuron. Given the current flow or potential at the surface of the neuron membrane, the problem is to calculate the potentials that would be measured in the surrounding tissue (which, for simplicity, is assumed to be a homogenous volume conductor). Under certain simplifying but not particularly unrealistic assumptions, it is permissible to replace the problem just stated with an equivalent and more familiar electrostatics problem. In fact, we may usually choose either of two different (but equivalent) electrostatic models for the actual situation. In the first, we replace an actual neuron by a model neuron of the same size and shape constructed from a nonconductor (plastic, for instance), embed this model neuron in a dielectric medium (air, for example), and place a dipole charge layer of a strength proportional to the membrane potential of the actual neuron on the model neuron's surface. The potential in the dielectric medium containing the model will then equal the potential in the tissue surrounding the actual neuron. In the second electrostatic model, everything is the same as in the model just described except that the surface of the model neuron is covered with a surface charge numerically equal to the normal component of the membrane current in the actual neuron. Again, the potential surrounding the model neuron is (approximately) equal to that in the tissue around the actual neuron. The advantage of these electrostatic models is that it is often fairly easy to predict qualitatively from them the potentials expected to surround an actual neuron. That is, one frequently can say in a certain situation whether the potential recorded at a point would be positive or negative, and whether it would be larger or smaller than the potential recorded at some other point. Such qualitative understanding is essential for a preliminary interpretation of the gross potentials recorded in the brain, and is often sufficiently precise for the purpose at hand. We shall now consider the justification for these conclusions.

The behavior of the brain's electromagnetic field is, of course, governed by Maxwell's equations

$$\nabla \cdot \mathbf{D} = 4\pi\rho \qquad \nabla \times \mathbf{H} = \frac{4\pi}{c}\mathbf{J} + \frac{1}{c}\frac{\partial \mathbf{D}}{\partial t}$$

$$\nabla \cdot \mathbf{B} = 0 \qquad \nabla \times \mathbf{E} = -\frac{1}{c}\frac{\partial \mathbf{B}}{\partial t}$$

Here, **E** is the electric field, **D** the dielectric displacement, **B** the magnetic flux density, **H** the magnetic field, c the speed of light, ρ the charge density, and **J** the current density. In addition to these basic equations we require the constitutive relation

$$\mathbf{J} = \sigma\mathbf{E} \qquad \text{(Ohm's law)}$$

As before, **J** is the electric current density and σ is the conductivity of the medium. We shall assume that σ is constant for the tissue surrounding the particular neuron on which we focus our attention; we have assumed the brain to be homogeneous and isotropic. Since virtually all of the brain volume not occupied with neurons is taken up by glia cells, the assumption of homogeneity is certain to be a crude one. Nevertheless, as long as we are not concerned with the behavior of the field over very small regions, this assumption is apparently adequate.

To further simplify the situation, we assume that the electromagnetic field is quasi-stationary. Specifically, this means that we consider only those situations in which $\partial\mathbf{D}/\partial t$ and $\partial\mathbf{B}/\partial t$ may be neglected. Although the fields resulting from PSPs are indeed slowly varying, the action potential itself has appreciable frequency components above ten thousand cycles per second, and it is thus stretching a point to consider these fields as quasi-stationary. Nevertheless, it is unlikely that errors introduced by this assumption will be of any practical importance in the typical neurophysiological application of the theory. With the assumption of quasi-stationarity, Maxwell's equations become

$$\nabla \cdot \mathbf{D} = 4\pi\rho \qquad \nabla \times \mathbf{H} = \frac{4\pi}{c}\mathbf{J}$$

$$\nabla \times \mathbf{E} = 0 \qquad \nabla \cdot \mathbf{B} = 0$$

Taking the divergence of the $\nabla \times \mathbf{H}$ relation, we have (remembering that div curl of a vector vanishes)

$$\nabla \cdot \mathbf{J} = 0$$

Substituting into this relation from Ohm's law, we obtain $\nabla \cdot \mathbf{E} = 0$. The behavior of the electric field in the brain is thus governed by the equation

$$\nabla \cdot \mathbf{E} = 0$$

$$\nabla \times \mathbf{E} = 0$$

These last relations are those for the electrostatic field in charge-free space, and form the basis for the electrostatic models mentioned earlier (page 162). In the remainder of the chapter we shall not further consider magnetic fields around a neuron, so that the two equations just given will serve alone as the basis for additional calculations.

As long as the preceding equations are limited to regions containing no electromotive forces, **E** is derivable from a scalar potential ϕ:

$$\mathbf{E} = -\nabla \phi$$

Substitution of this relation into $\nabla \cdot \mathbf{E} = 0$ yields the basic relation with which we shall carry out most of the manipulations to follow:

$$\nabla^2 \phi = 0 \qquad \text{(Laplace's equation, 10.27)}$$

We wish to obtain an integral representation of Laplace's equation. This is accomplished through Green's second identity, a relation holding for any scalar functions ϕ and G upon which the indicated operations can be legitimately carried out:

$$\int_V \phi \, \nabla^2 G \, dV = \int_V G \, \nabla^2 \phi \, dV - \oint_S \left(\phi \frac{\partial G}{\partial n} - G \frac{\partial \phi}{\partial n} \right) dS \qquad \text{(Green ii)}$$

Here $\partial/\partial n$ denotes the inward normal derivative. This theorem is the vector analytic version of ordinary integration by parts and may be obtained as follows. Into the divergence theorem

$$\int_V \nabla \cdot \mathbf{A} \, dV = \oint_S \mathbf{A} \cdot d\mathbf{S}$$

substitute $\phi \nabla G$ for **A**, remembering that $\nabla \cdot (\phi \nabla G) = \phi \nabla^2 G + \nabla \phi \cdot \nabla G$. If we denote $\phi \nabla G \cdot \mathbf{n}$ by $\phi (\partial G/\partial n)$ (**n** is the inward unit normal vector), this substitution yields Green's first identity:

$$\int_V (\phi \nabla^2 G + \nabla\phi \cdot \nabla G)\, dV = -\oint_S \phi \frac{\partial G}{\partial n}\, dS \qquad \text{(Green i)}$$

Writing Green i with ϕ and G interchanged, and subtracting from the noninterchanged Green i gives Green ii.

Green ii is a general relation which holds for any pair of functions ϕ and G. We may obtain an integral representation of Laplace's equation by identifying ϕ in Green ii with the ϕ in Laplace's equation (remembering that $\nabla^2\phi = 0$), and by letting $G = 1/|\mathbf{x} - \boldsymbol{\xi}|$:

$$\int_V \phi(\boldsymbol{\xi})\, \nabla^2 \frac{1}{|\mathbf{x} - \boldsymbol{\xi}|}\, d^3\boldsymbol{\xi} = -\oint_S \left(\phi(\boldsymbol{\xi}) \frac{\partial}{\partial n} \frac{1}{|\mathbf{x} - \boldsymbol{\xi}|} - \frac{1}{|\mathbf{x} - \boldsymbol{\xi}|} \frac{\partial\phi}{\partial n}\right) dS$$

The volume integral on the left can be further simplified by knowing that $\nabla^2\, 1/|\mathbf{x} - \boldsymbol{\xi}| = -4\pi\delta(\mathbf{x} - \boldsymbol{\xi})$, where $\delta(\mathbf{x} - \boldsymbol{\xi})$ is the three-dimensional Dirac delta function (see Jackson, p. 13). The integral representation for Laplace's equation then becomes

$$\phi(\mathbf{x}) = \frac{1}{4\pi} \oint_S \left(\phi(\boldsymbol{\xi}) \frac{\partial}{\partial n} \frac{1}{|\mathbf{x} - \boldsymbol{\xi}|} - \frac{1}{|\mathbf{x} - \boldsymbol{\xi}|} \frac{\partial\phi}{\partial n}\right) dS \qquad \text{if } \mathbf{x} \text{ is in the volume enclosed by } S.$$

$$\phi(\mathbf{x}) = 0 \qquad \text{if } \mathbf{x} \text{ is outside the volume enclosed by } S.$$

For our purposes, one final change in this equation will be useful. From Ohm's law we know that $\partial\phi/\partial n = -J_n/\sigma$ where J_n is the (inward) normal component of current at the surface. Finally then,

$$\phi(\mathbf{x}) = \frac{1}{4\pi} \oint \left(\phi(\boldsymbol{\xi}) \frac{\partial}{\partial n} \frac{1}{|\mathbf{x} - \boldsymbol{\xi}|} + \frac{1}{|\mathbf{x} - \boldsymbol{\xi}|} \frac{J_n}{\sigma}\right) dS \qquad \mathbf{x} \text{ in } V$$

$$\phi(\mathbf{x}) = 0 \qquad \mathbf{x} \text{ not in } V \tag{10.28}$$

Equation (10.28) is a general result found in discussions of electrostatics, and it is perhaps well to make its meaning explicit for the particular application to extracellular potentials. Under the assumptions we have made, the potential in the tissue surrounding a neuron must satisfy Laplace's equation. Through the manipulations indicated in the previous paragraph, we have transformed Laplace's equation into an integral representation in which the potential in the tissue surrounding a neuron may be calculated by performing an integration over the surface of the neuron.* To calculate the potential $\phi(\mathbf{x})$ at a point $\mathbf{x}$ outside of a particular neuron, we must know the surface potential $\phi(\mathbf{x}_{\text{surf}})$ of the neuron, and also the normal component of the membrane current at the surface $J_n(\mathbf{x}_{\text{surf}})$. Substituting these values into (10.28) yields, after integrating the potential and normal current over the surface as indicated, the desired potential. In principle, then, equation (10.28) completely solves any problem in which potentials surrounding a neuron must be calculated. Unfortunately, though, in its present form the equation is not of much practical use. The next question is how to recast equation (10.28) into a more tractable form.

Before continuing with manipulations on equation (10.28), however, we shall present the electrostatic model which is equivalent to the nonelectrostatic case just presented. Equation (10.28) relates the potential $\phi(\mathbf{x})$ (at point $\mathbf{x}$) in a volume conductor to the potential $\phi(\mathbf{x}_{\text{surf}})$ on the surface of the conductor (the neuronal surface) and the inward normal current $J_n(\mathbf{x}_{\text{surf}})$; exactly the same equation results if the volume conductor is replaced by a dielectric medium with dielectric constant 1, and if the surface of the medium (that is, the surface of the model neuron) is covered by a surface charge of density numerically equal to $\phi(\mathbf{x}_{\text{surf}})$ and a dipole layer of magnitude $J_n(\mathbf{x}_{\text{surf}})$. This is true because a surface dipole layer of magnitude $J_n(\mathbf{x}_{\text{surf}})$ (and a moment normal to the surface) produces a potential $\phi_{\text{dipole}}(\mathbf{x})$ according to

$$\phi_{\text{dipole}}(\mathbf{x}) = \oint_S \phi(\boldsymbol{\xi}) \frac{\partial}{\partial n} \frac{1}{|\mathbf{x} - \boldsymbol{\xi}|} \, dS$$

while a surface charge of density J_n/σ produces a potential $\phi_{\text{charge}}(\mathbf{x})$ at $\mathbf{x}$ given by (generalized Coulomb's law):

$$\phi_{\text{charge}}(\mathbf{x}) = \frac{1}{\sigma} \oint_S \frac{J_n(\boldsymbol{\xi})}{|\mathbf{x} - \boldsymbol{\xi}|} \, dS$$

* These relations hold only for a volume surrounded by closed surface and the volume we consider is the entire animal excluding the particular neuron under consideration; thus strictly speaking, we must carry out our surface integration over the neuron and also over the surface of the animal's body. However, the surface integral vanishes as the size of the surface increases, and so in practice we may neglect the integration over the animal's surface and instead integrate only over the neuronal surface.

The appropriate surface dipole layers and charges will produce, in a dielectric medium, then, exactly the same potential at each point as is produced in a conducting medium by a surface potential and a surface normal current. Since the electrostatic case is more familiar, we shall frequently make use of it in place of the actual (equivalent) volume conductor formulation.

We turn now to simplifications of equation (10.28). The first of these will be to eliminate the integral corresponding to the effect of a surface charge (that is, the integral involving the surface normal current), to obtain an integral involving what corresponds to a dipole layer on the neuronal surface. This dipole layer will have a magnitude (approximately) proportional to the membrane potential, and will thus yield a relation between the potential outside of a neuron and its membrane potential difference. Equation (10.28) gives the potential $\phi(\mathbf{x})$ at $\mathbf{x}$ due to *outside* surface potentials and currents

$$\phi(\mathbf{x}) = \frac{1}{4\pi}\oint_{S_0}\left(\phi_0(\boldsymbol{\xi})\frac{\partial}{\partial n}\frac{1}{|\mathbf{x}-\boldsymbol{\xi}|} + \frac{1}{|\mathbf{x}-\boldsymbol{\xi}|}\frac{J_{n0}(\boldsymbol{\xi})}{\sigma_0}\right)dS \qquad \mathbf{x} \text{ in } V_0 \tag{10.29}$$

$$\phi(\mathbf{x}) = 0 \qquad \mathbf{x} \text{ not in } V_0$$

where ϕ_0 is the outside surface potential, J_{n0} the outside surface normal current and σ_0 the outside conductivity; the integration is carried out over S_0 the outer surface of the neuron and V_0 denotes the volume outside of the neuron. This equation is identical to equation (10.28) except that we have placed subscripts on the appropriate functions to emphasize that they refer to the outer surface of the neuron. In other words, the outer potential is completely specified by the outside surface conditions, and no restrictions at all are placed on the conditions at the inner surface of the neuron membrane. However, exactly the same equation must also hold for inside the neuron

$$\phi(\mathbf{x}) = \frac{1}{4\pi}\oint_{S_i}\left(\phi_i(\boldsymbol{\xi})\frac{\partial}{\partial n}\frac{1}{|\mathbf{x}-\boldsymbol{\xi}|} + \frac{1}{|\mathbf{x}-\boldsymbol{\xi}|}\frac{J_{ni}(\boldsymbol{\xi})}{\sigma_i}\right)dS \qquad \mathbf{x} \text{ in } V_i$$

$$\phi(\mathbf{x}) = 0 \qquad \mathbf{x} \text{ not in } V_i$$

Here ϕ_i denotes the inside surface potential, J_{ni} the inside normal current and σ_i the inside conductivity; the integration is carried out over S_i, the inside surface of the neuron and V_i denotes the volume inside of the neuron. If the observation point $\mathbf{x}$ is outside the volume enclosed by the surface over which the integration is carried out, that is, if $\mathbf{x}$ is outside of the neuron, $\phi(\mathbf{x})$ vanishes and

$$\oint_{S_i} \phi_i(\boldsymbol{\xi}) \frac{\partial}{\partial n} \frac{1}{|\mathbf{x}-\boldsymbol{\xi}|}\, dS = -\oint_{S_i} \frac{1}{|\mathbf{x}-\boldsymbol{\xi}|} \frac{J_{ni}(\boldsymbol{\xi})}{\sigma_i}\, dS$$

Physically, this means the outside potential is determined by the outside boundary conditions and the inside potential by the inside boundary conditions.

Since the neuron membrane is only about 200 Å thick, the distance from the observation point $\mathbf{x}$ to a point on the outer surface is essentially the same as the distance to the corresponding point on the inner surface; this means that the preceding integrations may be carried out over the outer surface with only negligible error

$$\oint_{S_o} \phi_i(\boldsymbol{\xi}) \frac{\partial}{\partial n} \frac{1}{|\mathbf{x}-\boldsymbol{\xi}|}\, dS = -\frac{1}{\sigma_i} \oint_{S_o} \frac{J_{ni}(\boldsymbol{\xi})}{|\mathbf{x}-\boldsymbol{\xi}|}\, dS$$

Finally, it may be supposed that the inner normal current J_{ni} equals the outer normal current J_{no}. Thus, the equation becomes:

$$\oint_{S_o} \phi_i(\boldsymbol{\xi}) \frac{\partial}{\partial n} \frac{1}{|\mathbf{x}-\boldsymbol{\xi}|}\, dS = -\frac{1}{\sigma_i} \oint_{S_o} \frac{J_{no}(\boldsymbol{\xi})}{|\mathbf{x}-\boldsymbol{\xi}|}\, dS$$

Using this relation to eliminate the second surface integral in equation (10.29), we have finally

$$\phi(\mathbf{x}) = \frac{1}{4\pi} \oint_{S_o} \left(\phi_o(\boldsymbol{\xi}) - \frac{\sigma_i}{\sigma_o} \phi_i(\boldsymbol{\xi}) \right) \frac{\partial}{\partial n} \frac{1}{|\mathbf{x}-\boldsymbol{\xi}|}\, dS \quad (10.30)$$

This equation holds only for $\mathbf{x}$ outside of the neuron.

Equation (10.30) relates the potential at any point outside of a neuron (relative to a distant outside point) to the inner and outer surface potentials. Usually, we do not actually know the outer surface potential, but only the inside potential relative to the potential at a far distant point. That is, ϕ_i is the quantity directly measured in experiments, and is quite frequently taken as the value of the membrane potential (inside-outside surface potential difference) because outside potentials are often sufficiently small that they may be neglected relative to inside potentials. If ϕ_o is small compared to ϕ_i, and if inside and outside conductances are roughly comparable, the equation (10.30) becomes

$$\phi(\mathbf{x}) \approx -\frac{\sigma_i}{4\pi\sigma_o} \oint_{S_o} \phi_i(\boldsymbol{\xi}) \frac{\partial}{\partial n} \frac{1}{|\mathbf{x}-\boldsymbol{\xi}|} dS \qquad (10.31)$$

Here, then, is a comparatively simple relation between extracellular potentials and the experimentally accessible inside potential.

The electrostatic equivalent of equation (10.31) is instructive. If our model neuron is covered with a dipole layer whose magnitude at each point is proportional to the inside potential (or to the membrane potential, approximately), the electrostatic field around the model will equal the field around the actual neuron in the volume conductor. Using equation (10.31) it is possible to derive a number of results of use in neurophysiology. A typical type of argument is the following: we ask for the way in which extracellular potentials (associated, for example, with an action potential) decrease with distance from the neuron. The exact form of the relationship depends, of course, on the geometry of the particular neuron and the exact pattern of potential within the neuron. Nevertheless, because of the dipole-type source of the potential, an expansion in multipoles will contain only dipole and higher terms. This means that, unless the neuron is very highly symmetrical, the dipole term (which goes as the reciprocal of the square of the distance from the neuron) generally will dominate, since all other terms in the expansion involve reciprocal distance cubed or higher. In the immediate vicinity of the neuron membrane, then, the potential will approach the local value of ϕ_o, and at a distance will decrease with the reciprocal of the square of the distance (Figure 10-10).

Equation (10.30) or (10.31) greatly simplify the calculation of potential fields around a neuron, and provide a better intuitive understanding of extracellular potentials. An alternative approximate formulation, however, is traditional in neurophysiology. This second alternative eliminates the first surface integral in equation (10.29), leaving only the integral which corresponds, in the electrostatic model, to a surface charge on the neuronal surface. Since the remaining integral is simply a generalized form of

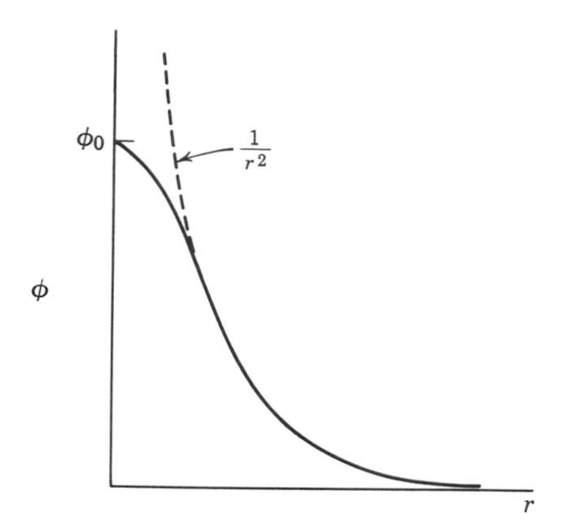

Figure 10-10. Magnitude of the extracellular potential ϕ as a function of distance r from a neuron.

Coulomb's law, intuitive understanding of the situation is further enhanced, and qualitative predictions are facilitated. Unfortunately, except for certain particular geometries, the errors inherent in the approximation are difficult to estimate. It is usually assumed, for a number of reasons, that nearly all of the membrane current is normal to the membrane at its surface. This means that J_n is approximately equal to the total membrane current i_m, and that the tangential component of the membrane current is small. To the extent that there is no tangential component to the current at the surface of the membrane, Ohm's law implies that the neuronal surface is isopotential. Taking ϕ outside the integral in equation (10.29), and performing the integration, one obtains ϕ for the first surface integral. Thus the contribution of this first integral to the potential is a constant, and, since $\phi(\mathbf{x})$ is arbitrary to an additive constant, it may be neglected to give

$$\phi(\mathbf{x}) = -\frac{1}{\sigma 4\pi} \oint_S \frac{i_m(\boldsymbol{\xi})}{|\mathbf{x} - \boldsymbol{\xi}|} \, dS \qquad (10.32)$$

The magnitude of the error committed by assuming isopotentiality of the neuron surface is unknown, but it certainly is not true that all membrane current is normal to the membrane. If the outside surface potential of the neuron is known, it is clear that the maximum error committed by (10.32) is less than the maximum difference in surface potential. This is not very helpful, however, since it only means that the error from using (10.32) is probably no larger than the value of the potential calculated.

The electrostatic model corresponding to equation (10.32) is particularly useful for making semi-quantitative predictions. If the model neuron is covered with a surface charge whose density at each point is equal in magnitude to the membrane current at that point (and whose sign is determined by the direction of the current, current flowing out of the neuron being positive), the potential in the dielectric surrounding the model will be given by (10.32). Membrane current itself is generally not known, but since in cylindrical structures like dendrites and axons it is given by the second partial of inside potential with respect to distance, it is often easy to calculate.

The use of equation (10.32), or rather, of the electrostatic model equivalent to it, may be indicated as follows: at a membrane where an action potential or an EPSP is occurring current flows into the neuron, drawing current out of the neighboring regions of membrane. At the site of an IPSP the opposite happens: current flows out of the active membrane and into neighboring regions. In terms of the electrostatic model, we would place negative charges at the site of an action potential or EPSP (putting positive charges on the neighboring surfaces), and place positive charges at the site of an IPSP (with an equal number of negative charges distributed over the neighboring regions). A region covered with positive charges is known as a **source,** and a negatively charged region is termed a **sink;** this terminology refers, of course, to the direction of current flow at the membrane surface. If a neurophysiologist records a positive potential in the brain, he concludes that he is nearer a source than a sink, and thus that an IPSP is occurring in the immediate vicinity of his electrode, or that an action potential or EPSP is occurring at a more remote region. A negative extracellular potential is given the opposite interpretation. By various experiments, it is possible to decide whether a particular potential results from an EPSP, an IPSP, or an action potential, and thus to make inferences about intracellular events from extracellular recordings. At the same time, this technique makes it possible to locate the relative position of synapses on a neuron: extracellular signs of EPSPs may be found in what is known to be one region of the neuron, while IPSPs can be seen to occur in a different region.

Useful as it is, the source-sink type of argument just described can also be very misleading if it is applied too literally. The difficulty can be illustrated by performing some calculations for a situation much like the one illustrated in Figure 10-9. Suppose that a delta function change in potassium and chloride conductances occurred at a single point ($y = 0$) along a very long (effectively infinite) dendrite to produce an IPSP at $t = 0$. Solving equation (10.27) for this situation, we obtain a solution * of the form

$$V(y, t) = \frac{1}{\sqrt{\pi t}}\, e^{(-y^2/4t)-t}$$

where $V(y, t)$ is the deviation of the inside potential from its resting value, y is the distance along the dendrite from the site of the IPSP (in units of the space constant; see p. 144) and t is the time after the conductance change (in time constant [p. 144] units). The second partial of $V(y, t)$ with respect to distance gives the membrane current, or equivalently the surface charge density in the case of the electrostatic model. Taking the derivative of membrane current with respect to distance, setting this equal to zero, and solving for y as a function t, we find the position y_{max} of the maximum inward current as a function of time

$$y_{max} = \sqrt{6t}$$

This equation thus gives the position of the maximum extracellular negative potential along the dendrite as a function of time. It is clear that the peak negativity propagates along the outer surface of the dendrite, and calculations show that velocities of about 0.1 or 1 meter per second might be obtained in an experiment. One (incorrect) interpretation of this moving extracellular negative

* This equation is most conveniently solved by taking the Laplace transform first with respect to time, and next with respect to distance (see Chapter 11 of Aseltine). The boundary and initial conditions are $V(y, o) = 0$ for all y, $V(\infty, t)$ vanishes for all t, and $\partial V/\partial y = 0$ for all t. This last condition follows from the fact that the deviation of the membrane potential from the resting value must be a maximum at all times at the site of origin of the IPSP.

potential would be that it is the extracellular sign of an action potential traveling along the dendrite with a velocity of about a meter per second. The correct interpretation, of course, is that the moving extracellular negativity is the consequence of an IPSP and the cable properties of the dendrite. Even this simple example illustrates the care that is needed when inferring intracellular events from extracellular potentials.

SUGGESTED REFERENCES

Chapter 1

Bloom, William and Don W. Fawcett: *A Textbook of Histology,* 8th ed. Philadelphia, W. B. Saunders Company, 1962. Chapter 9 presents a modern view of neuron and neuroglia structure and gives references to the standard textbooks of classical neuroanatomy.

Chapters 2 through 4

Ruch, Theodore C. and Harry D. Patton (editors): *Physiology and Biophysics,* 19th ed. Philadelphia, W. B. Saunders Company, 1965. The first half of this general physiology textbook is devoted to the nervous system and related topics. These sections constitute one of the standard surveys of neurophysiology at an intermediate level.

Bard, Philip (editor): *Medical Physiology,* 11th ed. St. Louis, C. V. Mosby Company, 1961. Another standard physiology textbook which covers much of the same material treated in Ruch and Patton, but often from a slightly different point of view.

Ochs, Sidney: *Elements of Neurophysiology.* New York, John Wiley & Sons, 1965. This general survey of neurophysiology is at a more advanced level than the present work, although not at quite as advanced a level as the two preceding references.

Eccles, John C.: *The Physiology of Nerve Cells.* Baltimore, The Johns Hopkins Press, 1957. A more specialized review of experiments on (for the most part) one particular type of neuron. Chapters describe the action potential, properties of the EPSP and of the IPSP. This book is familiar to all neurophysiologists, and the information contained in it serves as the conceptual framework for newer data.

Eccles, John C.: *The Physiology of Synapses.* New York, Academic Press, 1964. A more recent and already classic review of the properties of synapses. The structure of synapses, transmitters, mechanisms of the EPSP and IPSP, presynaptic inhibition, and many more topics are covered at an advanced level.

McLennan, Hugh: *Synaptic Transmission.* Philadelphia, W. B. Saunders Company, 1963. This monograph reviews the literature on the morphology, physiology, and pharmacology of synapses. Chapter 6 presents an extensive catalog of agents active at synapses.

Barron, D. H. and B. H. C. Matthews: The interpretation of potential changes in the spinal cord. *J. Physiol.*, **92**:276, 1938. Aside from containing a number of observations which have served as points of departure for many modern papers, this classic article presents what is probably the original statement of the slow potential theory.

Chapter 5

Granit, Ragnar: *Receptors and Sensory Perception.* New Haven, Yale University Press, 1955. Chapter 1 discusses the generator potential and its relation to the Weber-Fechner law.

Gray, J. A. B.: Initiation of impulses at receptors. In John Field (editor-in-chief): *Handbook of Physiology,* Section 1: Neurophysiology, Volume I. Washington, D.C., American Physiological Society, 1959. A more detailed and somewhat more specialized review of receptor physiology.

Bloom, William and Don W. Fawcett: *A Textbook of Histology,* 8th ed. Philadelphia, W. B. Saunders Company, 1962. Chapter 8 gives an excellent description of the structure of muscle. The physiology of muscle is well covered in Ruch and Patton or Bard (*op. cit.* Chapters 2 to 4 references).

Chapter 6

Hartline, H. K., Floyd Ratliff, and William H. Miller: Inhibitory interaction in the retina and its significance in vision. In E. Florey (editor): *Nervous Inhibition.* Oxford, Pergamon Press, 1961. This article is the definitive review of *Limulus* eye physiology, and also contains interpretations of contrast phenomena in terms of lateral inhibition.

Ratliff, Floyd: *Mach Bands: Quantitive studies on neural networks in the retina.* San Francisco, Holden-Day, 1965. This book is not only an excellent review of the inhibitory phenomena of the sort described in Chapter 5, but also relates the neurophysiology to broader philosophical questions in the theory of knowledge.

Chapter 7

Ehrlich, Annette: Neural control of feeding behavior. *Psychol. Bulletin,* **61**:100, 1964. This brief review article contains references to most of the important papers dealing with the hypothalamic "feeding centers."

Oomura, Y. and K. Kimura, et al.: Reciprocal activities of the ventromedial and lateral hypothalamic areas of cats. *Science,* **143:**484, 1964. A brief article reporting the results of experiments in which the behavior of single hypothalamic neurons was studied.

Chapter 8

Chow, K. L.: Anatomical and electrographical analysis of temporal neocortex in relation to visual discrimination learning in monkeys. In J. Delafresnaye (editor): *Brain Mechanisms and Learning.* Springfield, Charles C Thomas, 1961. This article reviews the evidence for storage sites in the brain for certain visual memories.

Sperry, R. W.: Cerebral organization and behavior. *Science,* **133:**1749, 1961. Review of work on the multiple storage of memories as revealed by the "split brain" technique.

Glickman, Stephen E.: Perservative neural processes and consolidation of the memory trace. *Psychol. Bulletin,* **58:**218, 1961. A review article which states the case in favor of the notion that a period of time is necessary for memories to be made permanent.

Skinner, B. F.: Are theories of learning necessary? *Psychol. Rev.,* **57:**193, 1950. This article contains a brief description of the pigeon experiment (described in the text) which demonstrates that time alone is not always sufficient to produce forgetting.

Bindman, Lynn J., O. C. J. Lippold and J. W. T. Redfearn: The action of brief polarizing currents on the cerebral cortex of the rat (1) during current flow and (2) in the production of long-lasting aftereffects. *J. Physiol.,* **172:**369, 1964. A paper describing one of the observations which might be related to the mechanism of information storage. Some of the results of this paper are mentioned briefly on page 112.

Dingman, Wesley and M. B. Sporn: Molecular theories of memory. *Science,* **144:**26, 1964. A critique of biochemical memory theories containing references to some of the more popular theories.

Chapters 9 and 10

Finkelstein, Alan and Alexander Mauro: Equivalent circuits as related to ionic systems. *Biophysical J.,* **3:**215, 1963. A discussion of different situations within a membrane (whose ionic fluxes are governed by the flux equation) leading to the equivalent circuit for the membrane given in Figure 10-1.

Cohen, H. and J. W. Cooley: The numerical solution of the time-dependent Nernst-Planck equations. *Biophysical J.,* **5:**145, 1965. Calculations described in this paper indicate that the steady-state ionic profiles within the neuron membrane should be attained in much less than a tenth of a microsecond.

Hodgkin, A. L. and W. A. H. Rushton: The electrical constants of a crustacean nerve fiber. *Proc. Roy. Soc. B. (London)*, **133:**444, 1946. A classic paper which applies the cable equation (equation [10.8]) to an axon.

Rall, Wilfrid: Electrophysiology of a dendritic neuron model. *Biophysical J.*, **2:**145, 1962. Application of the cable equation to the more complicated geometry of an idealized neuron. This article summarizes a number of earlier papers and contains references to them.

Cole, K. S. and H. J. Curtis: Electrical impedance of the squid giant axon during activity. *J. Gen. Physiol.*, **22:**649, 1939. This classic paper forms the foundation for the modern research on the action potential by demonstrating that it is membrane conductance that changes during the action potential.

Hodgkin, A. L. and A. F. Huxley: A quantitative description of membrane current and its application to conduction and excitation in nerve. *J. Physiol.*, **117:**500, 1952. The theoretical treatment of voltage clamp experiments presented in preceding articles (referenced in this article). This paper presents the Hodgkin-Huxley theory of the action potential and compares computed and recorded action potentials.

Hodgkin, A. L.: *The Conduction of the Nervous Impulse.* Springfield, Charles C Thomas, 1964. A review of data pertaining to the mechanism of the action potential by one of the major contributors to the field.

Araki, T. and C. A. Terzuolo: Membrane currents in spinal motoneurons associated with the action potential and synaptic activity. *J. Neurophysiol.*, **25:**772, 1962. Data presented in this paper indicate the time course of the conductance change associated with EPSPs and IPSPs in a particular type of neuron.

Lorente de No, R. A Study of Nerve Physiology. *Stud. Rockefeller Inst. Med. Research,* **132,** 1947. Chapter 16 of this treatise (often known as "the telephone book") contains the classic treatment of potential fields surrounding an axon bathed in a conducting medium.

Mauro, Alexander: Properties of thin generators pertaining to electrophysiological potentials in volume conductors. *J. Neurophysiol.*, **23:**132, 1960. This article presents the dipole model for potentials generated by an axon in a volume conductor, and derives a description of the extracellular fields associated with the passage of an action potential along an infinite axon.

Woodbury, J. Walter: Potentials in a volume conductor. In Chapter 3 of T. C. Ruch, and Harry D. Patton (editors): *Physiology and Biophysics,* 19 ed. Philadelphia, W. B. Saunders Company, 1965. A treatment of potential fields surrounding neurons in a volume conductor in terms of the dipole model. Several results are presented which aid in the qualitative application of the model.

Jackson, John D.: *Classical Electrodynamics.* New York, John Wiley & Sons, 1962. An excellent modern treatment of classical electricity and magnetism.

Aseltine, John A.: *Transform Method in Linear System Analysis.* New York, McGraw-Hill Book Company, 1958. A very readable treatment of the use of transforms for solving differential equations, including a chapter on solution of partial differential equations with the Laplace transform.

INDEX

Italicized numbers refer to figures

From page 41.

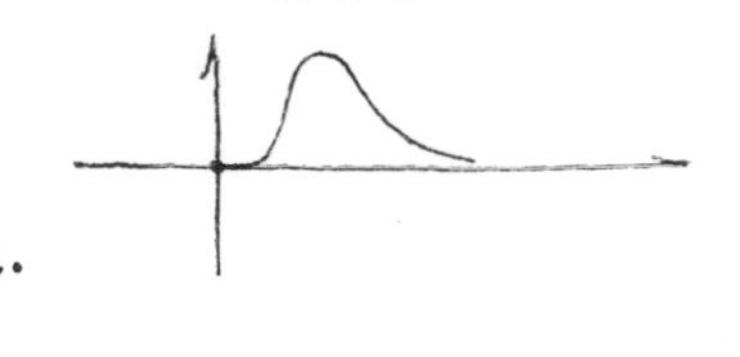

Single post-synaptic potential : θ.

1) θ is integrable, and non-negative.
2) θ is eventually decreasing.
3) θ vanishes on $(\leftarrow, 0]$.

Let α be a fixed positive number, representing the time interval between successive nerve (axonal) impulses.

$\mathbb{Z}_0^- = \{m \in \mathbb{Z} : m \leq 0$

We define:
$$\rho(t) = \Sigma_{m \in \mathbb{Z}_0^-} \theta(t+m\alpha)$$
$$\sigma(t) = \Sigma_{m \in \mathbb{Z}} \theta(t+m\alpha)$$
for each t in $\mathbb{R}$.

Since $\int_{\mathbb{R}} \theta(t)\,dt = \int_0^\alpha \sigma(t)\,dt$, we see that σ and ρ are finite almost everywhere. Clearly, σ is periodic of period α, and integrable on $[0,\alpha]$.

For each ε in $\mathbb{R}^+$, we may select T in $\mathbb{R}$ so that θ is decreasing on $[T,\rightarrow)$ and so that, for each t in $[T,\rightarrow)$, $\int_t^\infty \theta(t)\,dt < \alpha\varepsilon$. Then, for each t in $[T,\rightarrow)$, $\sigma(t) - \rho(t)$ $= \Sigma_{m \in \mathbb{Z}^+} \theta(t+m\alpha) \leq \frac{1}{\alpha} \int_t^\infty \theta(t)\,dt < \varepsilon$. Hence, ρ is (eventually) nearly periodic.

The time averages $\bar{\rho}$ and $\bar{\sigma}$ (that is, $\lim_{t\uparrow\infty} \frac{1}{t} \int_0^t \rho(t)\,dt$ and $\lim_{t\uparrow\infty} \frac{1}{t} \int_0^t \sigma(t)\,dt$) are clearly equal, and the latter (in virtue of periodicity) is $\frac{1}{\alpha}\int_0^\alpha \sigma(t)\,dt$. So then: $\bar{\rho} = \frac{1}{\alpha}\int_{\mathbb{R}} \theta(t)\,dt$.